T0297927

An Exponential Function Approach to Parabolic Equations

SERIES ON CONCRETE AND APPLICABLE MATHEMATICS

ISSN: 1793-1142

Series Editor: Professor George A. Anastassiou
Department of Mathematical Sciences
The University of Memphis
Memphis, TN 38152, USA

Published*

*To view the complete list of the published volumes in the series, please visit:
http://www.worldscientific/series/scaam

Series on Concrete and Applicable Mathematics – Vol.15

An Exponential Function Approach to Parabolic Equations

Chin-Yuan Lin

National Central University, Taiwan

 **World Scientific**

NEW JERSEY · LONDON · SINGAPORE · BEIJING · SHANGHAI · HONG KONG · TAIPEI · CHENNAI

Published by

World Scientific Publishing Co. Pte. Ltd.

5 Toh Tuck Link, Singapore 596224

USA office: 27 Warren Street, Suite 401-402, Hackensack, NJ 07601

UK office: 57 Shelton Street, Covent Garden, London WC2H 9HE

British Library Cataloguing-in-Publication Data
A catalogue record for this book is available from the British Library.

Series on Concrete and Applicable Mathematics — Vol. 15
AN EXPONENTIAL FUNCTION APPROACH TO PARABOLIC EQUATIONS

Copyright © 2015 by World Scientific Publishing Co. Pte. Ltd.

ISBN 978-981-4616-38-6

Printed in Singapore

To my wife and brothers,

and

in memory of my mother, Liu Gim

Preface

This book is intended for graduate students of mathematics, as well as for researchers. It requires a basic knowledge of both introductory functional analysis and elliptic partial differential equations of second order. However, for the convenience of the reader, some essential background results from elliptic partial differential equations are collected in the Appendix.

As is revealed by its contents, the book consists of nine chapters. Chapter 1 is on the existence of exponential functions of linear or nonlinear, time-independent operators, and Chapter 2 is on the same subject but with time-dependent operators. Chapters 3 through 6 are on applications of Chapters 1 and 2 to initial-boundary value problems for parabolic partial differential equations. Chapter 7 is on the associated elliptic equations with Chapters 3 through 6. Chapter 8 is on an extension of Chapter 2, whose results are applied to solve more general initial-boundary value problems in Chapter 9.

The author is indebted to Professors Jerome A. Goldstein and Gisele R. Goldstein at the University of Memphis, and to the late Professor Sen-Yen Shaw from the National Central University, for their teaching.

Finally, the author wishes to thank his wife and son for their support and encouragement.

Contents

CHAPTER 1

Existence Theorems for Cauchy Problems

1. Introduction

In this chapter, linear and nonlinear Cauchy problems, together with their associated nonhomogeneous problems, will be studied. Those problems will be solved with the aid of elementary difference equations. The obtained results will be illustrated by solving simple, initial-boundary value problems for parabolic, partial differential equations with time-independent coefficients. Further illustrations of solving more general, parabolic partial differential equations with time-independent coefficients will be given in Chapters 3 and 4.

Let constants $\omega \in \mathbb{R}$ and $M \geq 1$. Consider the linear Cauchy problem

$$\frac{d}{dt}u(t) = Bu(t), \quad t > 0$$
$$u(0) = u_0 \tag{1.1}$$

in a real Banach space $(X, \|\cdot\|)$, where u is a function from $[0, \infty)$ to X, and

$$B : D(B) \subset X \longrightarrow X$$

is an unbounded linear operator. Here recall

- A real Banach space is a complete, real normed vector space equipped with a norm.

  For example, the real vector space $\mathbb{R}$ of real numbers over the field of itself, equipped with the norm of the usual function $|\cdot|$ of absolute value, is a real Banach space.

  Another example is the real Banach space $(C[0, 1], \|\cdot\|_\infty)$ of all continuous, real-valued functions on $[0, 1]$, equipped with supremum norm $\|\cdot\|_\infty$, where

$$\|f\|_\infty \equiv \sup_{x \in [0,1]} |f(x)| \quad \text{if } f \in C[0, 1].$$

- One example of an unbounded linear operator is the first order ordinary differential operator S in the real Banch space $(C[0, 1], \|\cdot\|_\infty)$, where

$$S : D(S) \subset C[0, 1] \longrightarrow C[0, 1],$$

  defined by $(Sf)(x) \equiv \frac{d}{dx}f(x)$ for f in $D(S)$, the set of all real, continuously differentiable functions on $[0, 1]$.

To solve the linear Cauchy problem (1.1), let the simple case be considered first where $X = \mathbb{R}$, and $B = b$, a real number. In this case, the unique solution is given by

$$u(t) = e^{tb}u_0,$$

where the exponential function e^{tb} can be represented by this limit

$$e^{tb} = \lim_{n \to \infty} (1 - \frac{t}{n}b)^{-n}.$$

Thus the Cauchy problem (1.1) might be solved, if the quantity

$$(I - \frac{t}{n}B)^{-n}u_0$$

can be defined for each $t > 0$ and for each $n = 1, 2, \ldots$, and has a limit as $n \longrightarrow \infty$. Here I is the identity operator. But this will be true under suitable assumptions on the operator B, as the following describes it.

The linear Cauchy problem (1.1) will be solved under the assumption that the operator B satisfies both the range condition (B1) and the dissipativity condition (B2) or satisfies, more generally, the mixture condition (B3).

(B1) For each small $0 < \lambda < \lambda_0$, where λ_0 is some positive number satisfying $\lambda_0 \omega < 1$, the range of $(I - \lambda B)$ contains the closure $\overline{D(B)}$ of $D(B)$.

(B2) For each $v \in D(B)$, the inequality

$$\|v\| \leq \|v - \lambda(B - \omega)v\|$$

is true for each $\lambda > 0$.

(B3) For each $x \in \overline{D(B)}$ and for each small $0 < \lambda < \lambda_0$, where λ_0 is some positive number satisfying $\lambda_0 \omega < 1$, the quantity $(I - \lambda B)^{-1}x$ is single-valued, and the inequality is true:

$$\|(I - \lambda B)^{-n}x\| \leq M(1 - \lambda\omega)^{-n}\|x\|, \quad n = 1, 2, \ldots.$$

Here I is the identity operator.

Under both the (B1) and the (B2) or under, more generally, the (B3), the quantity

$$(I - \lambda B)^{-n}v, \quad n = 1, 2, \ldots$$

will be well-defined for each small $\lambda > 0$ satisfying $\lambda\omega < 1$ and for each $v \in \overline{D(B)}$.

It is the first of our three purposes in this chapter to show, using the difference equations theory [28], that the limit

$$\lim_{n \to \infty} (I - \frac{t}{n}B)^{-n}v$$

exists for bounded $t \geq 0$. From this, it will follow that, if u_0 lies in $D(B)$ with $Bu_0 \in \overline{D(B)}$ and if B is additionally a closed operator, then

$$u(t) \equiv \lim_{n \to \infty} (I - \frac{t}{n}B)^{-n}u_0$$

is the unique solution of the linear Cauchy problem (1.1), in the sense that $u(t)$ is the unique continuously differentiable function of $t > 0$ satisfying (1.1). Here B is said to be a closed operator, if the condition

$$x_n \longrightarrow x \text{ and } Bx_n \longrightarrow y \quad \text{for } x_n \in D(B)$$

implies

$$x \in D(B) \text{ and } y = Bx.$$

For example, the first order ordinary differential operator S, mentioned above, is a closed operator.

The second purpose of this chapter is to study the nonlinear analogue of (1.1), the nonlinear Cauchy problem in the real Banach space X

$$\frac{d}{dt}u(t) \in Au(t), \quad t > 0$$

$$u(0) = u_0,$$

(1.2)

where $A : D(A) \subset X \longrightarrow X$, is a nonlinear, multi-valued operator with domain $D(A)$. The $D(A)$ is not necessarily a subspace of X, a case different from that with the linear Cauchy problem (1.1), and one example of such an A is the nonlinear, multi-valued function

$$f : D(f) \subset \mathbb{R} \longrightarrow \mathbb{R},$$

defined by

$$f(x) = \begin{cases} x^2 + 1, & \text{if } 0 < x < 2; \\ [-1, 1], & \text{if } x = 0; \\ -x^2 - 1, & \text{if } -2 < x < 0; \end{cases}$$

in the real Banach space $\mathbb{R}$. Here the domain $D(f) = (-2, 2)$ of f is not a subspace of $\mathbb{R}$, and the function value $f(0)$ of f at $x = 0$ is not a single number but a set $[-1, 1]$.

To solve the nonlinear Cauchy problem (1.2), let its approximate problem, a difference equation, be examined first [**14**, Page 9], [**8**, Page 72]

$$\frac{u_\epsilon(t) - u_\epsilon(t - \epsilon)}{\epsilon} \in Au_\epsilon(t), \quad t > 0,$$

$$u_\epsilon(0) = u_0.$$

(1.3)

Here $\epsilon > 0$ is very small. This approximate problem (1.3) becomes the equation

$$u_\epsilon(t) = [I - \epsilon A]^{-1} u_\epsilon(t - \epsilon), \quad t > 0,$$

$$u_\epsilon(0) = u_0,$$

(1.4)

if the quantity

$$(I - \epsilon A)^{-1}$$

can be defined. Here I is the identity operator. Thus when $\epsilon = \frac{t}{n}$ for $n \in \mathbb{N}$, the equation (1.4) is readily seen to have the solution

$$u_\epsilon(t) = (I - \epsilon A)^{-1}(I - \epsilon A)^{-1} \cdots (I - \epsilon A)^{-1} u_0$$

$$\equiv (I - \epsilon A)^{-n} u_0,$$

provided that the quantity

$$(I - \epsilon A)^{-n} u_0$$

can be defined for each $t > 0$ and for each $n = 1, 2, \ldots$. Therefore, the nonlinear Cauchy problem (1.2) might be solved, if the limit exists, as $n \longrightarrow \infty$, of the quantity

$$(I - \epsilon A)^{-n} u_0 = (I - \frac{t}{n}A)^{-n} u_0.$$

But this will be true under suitable assumptions on the nonlinear operator A, as the following describes it.

The nonlinear Cauchy problem will be solved under the similar assumption that A satisfies both the range condition (A1) and the dissipativity condition (A2).

(A1) For each small $0 < \lambda < \lambda_0$, where λ_0 is some positive number satisfying $\lambda_0 \omega < 1$, the range of $(I - \lambda A)$ contains the closure $\overline{D(A)}$ of $D(A)$.

(A2) For each $v, w \in D(A)$, the inequality

$$\|v - w\| \le \|(v - w) - \lambda(x - y)\|$$

is true for each $\lambda > 0$ and for all $x \in (A - \omega)v, y \in (A - \omega)w$.

As is the case with B, the quantity

$$(I - \lambda A)^{-n}v, \quad n = 1, 2, \ldots,$$

under the conditions (A1) and (A2), will be well-defined for each small $\lambda > 0$ satisfying $\lambda\omega < 1$ and for each $v \in \overline{D(A)}$. Using the difference equations theory again [28], it will be shown that the limit

$$\lim_{n\to\infty} (I - \frac{t}{n}A)^{-n}v$$

exists for bounded $t \ge 0$. Different from the case with the linear Cauchy problem (1.1), the quantity

$$u(t) \equiv \lim_{n\to\infty} (I - \frac{t}{n}A)^{-n}u_0$$

for $u_0 \in D(A)$ will be only interpreted as a limit solution of the nonlinear Cauchy problem (1.2). However, $u(t)$ will be a strong solution if A is what we call embeddedly quasi-demi-closed; this will be explained in Section 2.

The third (final) purpose of this chapter is to study the simple nonhomogeneous equations associated with (1.1) and (1.2), respectively:

$$\frac{d}{dt}u(t) = Bu + f_0, \quad t > 0$$
$$u(0) = u_0$$

(1.5)

and

$$\frac{d}{dt}u(t) \in Au + f_1, \quad t > 0$$
$$u(0) = u_0.$$

(1.6)

Here f_0 and f_1 are two elements in X.

The results of this chapter will be used in Chapters 3 and 4, where the corresponding operators B or A are second order, elliptic differential operators, and the corresponding Cauchy problems are parabolic, initial-boundary value problems. Since the quantities

$$\lim_{n\to\infty} (I - \frac{t}{n}B)^{-n}u_0 \text{ and } \lim_{n\to\infty} (I - \frac{t}{n}A)^{-n}u_0$$

are similar to the ordinary, real-valued, exponential function of t

$$e^{ta} = \lim_{n\to\infty} (1 - \frac{t}{n}a)^{-n}, \quad a \in \mathbb{R},$$

the title of this book is explained.

Finally, we organize the rest of this chapter as follows, which consists of five other sections. Section 2 states the main results, and Section 3 illustrates the main results by simple examples. Section 4 gives some preliminary results, and Section 5 presents a basic theory of difference equations. The last section, Section 6, proves the main results, using the results in Sections 4 and 5.

2. Main Results

2.1. Linear Homogeneous Equations. With regard to the linear Cauchy problem (1.1), we have Theorems 2.1, 2.2, and 2.3.

THEOREM 2.1 (A classical solution). *Let the linear unbounded operator B satisfy both the range condition (B1) and the dissipativity condition (B2) or satisfy, more generally, the mixture condition (B3). Then the limit*

$$S(t)x \equiv \lim_{n \to \infty} (I - \frac{t}{n}B)^{-n}x$$
$$= \lim_{\nu \to 0} (I - \nu B)^{-[\frac{t}{\nu}]}x$$

exists for each $x \in \overline{D(B)}$ and for bounded $t \geq 0$.

This limit $S(t)x$ is also continuous in $t \geq 0$ for $x \in \overline{D(B)}$, but Lipschitz continuous in $t \geq 0$ for $x \in D(B)$.

Furthermore, if u_0 lies in $D(B)$ with $Bu_0 \in \overline{D(B)}$, and if B is additionally a closed operator, then the function

$$u(t) \equiv S(t)u_0 = \lim_{n \to \infty} (I - \frac{t}{n}B)^{-n}u_0$$
$$= \lim_{\nu \to 0} (I - \nu B)^{-[\frac{t}{\nu}]}u_0$$

is the unique solution of the linear Cauchy problem (1.1).

THEOREM 2.2 (Regularity of solution). *Following Theorem 2.1,*

$$\frac{d}{dt}u(t) = Bu(t) = BS(t)u_0$$
$$= S(t)(Bu_0)$$

is continuous in $t \geq 0$ for $u_0 \in D(B)$ with $Bu_0 \in \overline{D(B)}$, is Lipschitz continuous in $t \geq 0$ for $u_0 \in D(B^2)$, and is differentiable in $t \geq 0$ for $u_0 \in D(B^2)$ with $B^2u_0 \in \overline{D(B)}$. More regularity of $\frac{d}{dt}u(t)$ in t can be obtained iteratively.

From the proofs of Theorems 2.1 and 2.2 in Subsection 6.2, it follows readily that

THEOREM 2.3. *The results in Theorems 2.1 and 2.2 are still true, if the range condition (B1) and the mixture condition (B3) are replaced by the weaker conditions (B1)′ and (B3)′ below, respectively, provided that the initial conditions $x \in \overline{D(B)}$, $Bu_0 \in \overline{D(B)}$, and $B^2u_0 \in \overline{D(B)}$, are changed to the initial conditions $x \in D(B)$, $Bu_0 \in D(B)$, and $B^2u_0 \in D(B)$, respectively. Here*

(B1)′ *For each small $0 < \lambda < \lambda_0$, where λ_0 is some positive number satisfying $\lambda_0\omega < 1$, the range of $(I - \lambda B)$ contains $D(B)$.*

(B3)′ *For each $x \in D(B)$ and for each small $0 < \lambda < \lambda_0$, where λ_0 is some positive number satisfying $\lambda_0\omega < 1$, the quantity $(I - \lambda B)^{-1}x$ is single-valued, and the inequality is true:*

$$\|(I - \lambda B)^{-n}x\| \leq M(1 - \lambda\omega)^{-n}\|x\|, \quad n = 1, 2, \ldots.$$

Remark. Theorems 2.1 and 2.2 are the Hille-Yosida Theorem [9, 10, 14, 16, 17, 30, 31, 39], if the linear operator B is densely defined and satisfies the dissipativity condition (B2) and the stronger range condition (B1)″:

$(B1)''$ For each small $0 < \lambda < \lambda_0$, where λ_0 is some positive number satisfying $\lambda_0 \omega < 1$, the range of $(I - \lambda B)$ equals X.

In the present case, the section $B^{\overline{D(B)}}$ of B on the Banach space $\overline{D(B)}$,

$$B^{\overline{D(B)}} : D(B) \subset \overline{D(B)} \longrightarrow \overline{D(B)},$$

is not necessarily well-defined, as $B^{\overline{D(B)}}x = Bx$ for $x \in D(B)$ may lie outside of $\overline{D(B)}$. Therefore, Theorems 2.1 and 2.2 will not follow from applying the Hille-Yosida theorem to the section $B^{\overline{D(B)}}$.

2.2. Nonlinear Homogeneous Equations. With regard to the nonlinear Cauchy problem (1.2), we have Theorems 2.4, 2.5, and 2.6.

THEOREM 2.4 ((Existence of a limit) Crandall-Liggett theorem [**6, 30**]). *Let the nonlinear, multi-valued operator A satisfy both the range condition (A1) and the dissipativity condition (A2). Then the limit*

$$U(t)x \equiv \lim_{n \to \infty} \left(I - \frac{t}{n}A\right)^{-n} x$$

$$= \lim_{\nu \to 0} (I - \nu A)^{-\left[\frac{t}{\nu}\right]} x$$

exists for each $x \in \overline{D(A)}$ and for bounded $t \geq 0$.

This limit $U(t)x$ is also continuous in $t \geq 0$ for $x \in \overline{D(A)}$, but Lipschitz continuous in $t \geq 0$ for $x \in D(A)$.

In order to state the next theorem, Theorem 2.5, concerning a limit solution and a strong solution, we need to make two preparations.

The first preparation is for a limit solution. Let $T > 0, u_0 \in D(A)$, and $n \in \mathbb{N}$ be large. Consider the discretization of (1.2) on $[0, T]$

$$u_i - \nu A u_i \ni u_{i-1}, u_i \in D(A), \qquad (2.1)$$

where $\nu = \frac{T}{n}$ satisfies $\nu < \lambda_0$ for which

$$\nu \omega < 1, \quad i = 1, 2, \dots, n,$$

and u_i will exist uniquely by the range condition (A1) and the dissipativity condition (A2) (see Section 4).

Putting $t_i = i\nu$ and defining the Rothe functions [**12, 32**]

$$\chi^n(0) = u_0;$$
$$\chi^n(t) = u_i \quad \text{for } t \in (t_{i-1}, t_i];$$
$$u^n(t) = u_{i-1} + \frac{u_i - u_{i-1}}{\nu}(t - t_{i-1}) \qquad (2.2)$$
$$\text{for } t \in [t_{i-1}, t_i],$$

it will follow (see Section 6) that

$$\limsup_{n \to \infty} \|u^n(t) - \chi^n(t)\| = 0;$$

$$\frac{du^n(t)}{dt} \in A\chi^n(t), u^n(0) = u_0 \qquad (2.3)$$

$$\text{for almost every } t.$$

Here the last equation has values in $B([0, T]; X)$, the real Banach space of bounded functions from $[0, T]$ to X.

The other preparation is for a strong solution. Let $(Y, \|.\|_Y)$ be a real Banach space, into which the real Banach space $(X, \|.\|)$ is continuously embedded. For example, the space $(C[0,1], \| \cdot \|_\infty)$ is continuously embedded into the real Banach space $(L^2(0,1), \| \cdot \|_2)$ of Lebesgue square integrable functions on $(0,1)$ by the identity mapping.

The nonlinear, multi-valued operator A is said to be embeddedly quasi-demi-closed with respect to X and Y, if it satisfies the embedding condition (A3).

(A3) If $x_n \in D(A) \longrightarrow x$ and if $\|y_n\| \leq M_0$ for some $y_n \in Ax_n$ and for some positive constant M_0, then $x \in D(\eta \circ A)$, the domain of $\eta \circ A$, (that is, $\eta(A(x))$ exists), and

$$|\eta(y_{n_l}) - z| \longrightarrow 0$$

for some subsequence y_{n_l} of y_n, for some $z \in \eta(Ax)$, and for each $\eta \in Y^* \subset X^*$, the real dual space of Y.

THEOREM 2.5 ((A limit or strong solution) [20]). *Following Theorem 2.4, if $u_0 \in D(A)$, then the function*

$$u(t) \equiv U(t)u_0 = \lim_{n \to \infty} (I - \frac{t}{n}A)^{-n}u_0$$
$$= \lim_{\nu \to 0}(I - \nu A)^{-[\frac{t}{\nu}]}u_0$$

is a limit solution of the nonlinear Cauchy problem (1.2) on $[0,T]$ and then on $[0,\infty)$, in the sense that it is also the uniform limit of $u^n(t)$ on $[0,T]$, where $u^n(t)$ satisfies (2.3) on $[0,T]$ and $T > 0$ is arbitrary.

Furthermore, if A is embeddedly quasi-demi-closed, that is, if A satisfies the embedding condition (A3), then $u(t)$ is a strong solution of (1.2) in Y, that is, $u(t)$ satisfies (1.2) in Y for almost every t.

It follows readily from the proofs of Theorems 2.4 and 2.5 in Subsection 6.1 that

THEOREM 2.6. *The results in Theorems 2.4 and 2.5 are still true, if the range condition (A1) is replaced by the weaker range condition (A1)' below, provided that the initial condition $x \in \overline{D(A)}$ is changed to the condition $x \in D(A)$. Here*

(A1)' *For each samll $0 < \lambda < \lambda_0$, where λ_0 is some positive number satisfying $\lambda_0\omega < 1$, the range of $(I - \lambda A)$ contains $D(A)$.*

2.3. Linear Nonhomogeneous Equations. With regard to the nonhomogeneous, linear Cauchy problem (1.5), we have Theorems 2.7, 2.8, and 2.9. Here the element f_0 in (1.5) will be conditioned by (F0):

(F0) $(y + \lambda f_0)$ lies in the range of $(I - \lambda B)$ whenever $y \in \overline{D(B)}$ and

$$0 < \lambda < \lambda_0 \quad \text{with } \lambda_0\omega < 1.$$

It is readily seen that, if $f_0 \in \overline{D(B)}$, then f_0 satisfies the condition (F0).

THEOREM 2.7 (A classical solution). *Let a nonlinear operator*

$$\tilde{B} : D(\tilde{B}) \subset X \longrightarrow X$$

be defined by

$$\tilde{B}v = Bv + f_0$$

for $v \in D(\tilde{B}) \equiv D(B)$, where B is the B in Theorem 2.1, and $f_0 \in X$ satisfies the condition (F0).

Then the limit

$$\tilde{S}(t)x \equiv \lim_{n \to \infty} (I - \frac{t}{n}\tilde{B})^{-n}x$$

$$= \lim_{\nu \to 0} (I - \nu\tilde{B})^{-[\frac{t}{\nu}]}x$$

exists for each $x \in \overline{D(B)}$ and for bounded $t \geq 0$. Here $\tilde{B}$ need not satisfy both (A1) and (A2), but it does so, if B is dissipative.

This limit $\tilde{S}(t)x$ is also continuous in $t \geq 0$ for $x \in \overline{D(B)}$, but Lipschitz continuous in $t \geq 0$ for $x \in D(B)$.

Furthermore, if u_0 lies in $D(B)$ with $Bu_0, f_0 \in \overline{D(B)}$, and if B is additionally a closed operator, then the function

$$\tilde{u}(t) \equiv \tilde{S}(t)u_0 = \lim_{n \to \infty} (I - \frac{t}{n}\tilde{B})^{-n}u_0$$

$$= \lim_{\nu \to 0} (I - \nu\tilde{B})^{-[\frac{t}{\nu}]}u_0$$

is the unique solution of the nonhomogeneous, linear Cauchy problem (1.5).

THEOREM 2.8 (Regularity of solution). *Follow Theorem 2.7 and use the $S(t)$ in Theorem 2.1. It follows that*

$$\frac{d}{dt}\tilde{u}(t) = \tilde{B}\tilde{u}(t) = \tilde{B}\tilde{S}(t)u_0$$

$$= S(t)(\tilde{B}u_0)$$

$$= S(t)(Bu_0 + f_0)$$

is continuous in $t \geq 0$ for $u_0 \in D(B)$ with $Bu_0, f_0 \in \overline{D(B)}$, is Lipschitz continuous in $t \geq 0$ for $u_0 \in D(B^2)$ and for $f_0 \in D(B)$, and is differentiable in $t \geq 0$ for $u_0 \in D(B^2)$ and for $f_0 \in D(B)$ with $B^2u_0, Bf_0 \in \overline{D(B)}$. More regularity of $\frac{d}{dt}\tilde{u}(t)$ in t can be obtained iteratively.

From the proofs of Theorems 2.7 and 2.8 in Subsection 6.3, it follows readily that

THEOREM 2.9. *The results in Theorems 2.7 and 2.8 are still true, if the range condition (B1), the mixture condition (B3), and the condition (F0) are replaced by the weaker conditions (B1)′, (B3)′, and (F0)′ below, respectively, provided that the initial conditions $x \in \overline{D(B)}, Bu_0 \in \overline{D(B)}$, and $B^2u_0 \in \overline{D(B)}$ and the conditions $f_0 \in \overline{D(B)}$ and $Bf_0 \in \overline{D(B)}$ are changed to the initial conditions $x \in D(B), Bu_0 \in D(B)$, and $B^2u_0 \in D(B)$ and the conditions $f_0 \in D(B)$ and $Bf_0 \in D(B)$, respectively. Here*

(B1)′ *For each small $0 < \lambda < \lambda_0$, where λ_0 is some positive number satisfying $\lambda_0\omega < 1$, the range of $(I - \lambda B)$ contains $D(B)$.*

(B3)′ *For each $x \in D(B)$ and for each small $0 < \lambda < \lambda_0$, where λ_0 is some positive number satisfying $\lambda_0\omega < 1$, the quantity $(I - \lambda B)^{-1}x$ is single-valued, and the inequality is true:*

$$\|(I - \lambda B)^{-n}x\| \leq M(1 - \lambda\omega)^{-n}\|x\|, \quad n = 1, 2, \ldots.$$

$(F0)'$ $(y + \lambda f_0)$ *lies in the range of* $(I - \lambda B)$ *whenever* $y \in D(B)$ *and*

$$0 < \lambda < \lambda_0 \quad \text{with } \lambda_0 \omega < 1.$$

2.4. Nonlinear Nonhomogeneous Equations. With regard to the nonhomogeneous, nonlinear Cauchy problem (1.6), we have Theorems 2.10, 2.11, and 2.12. Here the element f_1 in (1.6) will be conditioned by either (F1) or weaker (F2):

(F1) $(y + \lambda f_1)$ *lies in the range of* $(I - \lambda A)$ *whenever* $y \in \overline{D(A)}$ *and*

$$0 < \lambda < \lambda_0 \quad \text{with } \lambda_0 \omega < 1.$$

(F2) $(y + \lambda f_1)$ *lies in the range of* $(I - \lambda A)$ *whenever* $y \in D(A)$ *and*

$$0 < \lambda < \lambda_0 \quad \text{with } \lambda_0 \omega < 1.$$

THEOREM 2.10 (Existence of a limit). *Let a nonlinear operator* $\tilde{A} : D(\tilde{A}) \subset X \longrightarrow X$ *be defined by*

$$\tilde{A}v = Av + f_1$$

for $v \in D(\tilde{A}) \equiv D(A)$, *where* A *is the* A *in Theorem 2.4, and* $f_1 \in X$ *satisfies the condition (F1).*

Then the limit

$$\tilde{U}(t)x \equiv \lim_{n \to \infty} (I - \frac{t}{n}\tilde{A})^{-n}x$$
$$= \lim_{\nu \to 0}(I - \nu\tilde{A})^{-[\frac{t}{\nu}]}x$$

exists for each $x \in \overline{D(A)}$ *and for bounded* $t \geq 0$. *Here* $\tilde{A}$ *satisfies both (A1) and (A2).*

This limit $\tilde{U}(t)x$ *is also continuous in* $t \geq 0$ *for each* $x \in \overline{D(A)}$, *but Lipschitz continuous in* $t \geq 0$ *for* $x \in D(A)$.

THEOREM 2.11 (A limit or strong solution). *Following Theorem 2.10, if* $u_0 \in D(A)$, *then the function*

$$\tilde{u}(t) \equiv \tilde{U}(t)u_0 = \lim_{n \to \infty} (I - \frac{t}{n}\tilde{A})^{-n}u_0$$
$$= \lim_{\nu \to 0}(I - \nu\tilde{A})^{-[\frac{t}{\nu}]}u_0$$

is a limit solution of the nonlinear Cauchy problem (1.6) on $[0, T]$ *and then on* $[0, \infty)$, *in the sense that it is also the uniform limit of* $u^n(t)$ *on* $[0, T]$, *where* $u^n(t)$ *satisfies (2.3) on* $[0, T]$ *with* A *replaced by* $\tilde{A}$, *and* $T > 0$ *is arbitrary.*

Furthermore, if A *is embeddedly quasi-demi-closed, that is, if* A *satisfies the embedding condition (A3), then* $\tilde{u}(t)$ *is a strong solution of (1.6) in* Y, *that is,* $\tilde{u}(t)$ *satisfies (1.6) in* Y *for almost every* t.

It follows readily from the proofs of Theorems 2.10 and 2.11 in Subsection 6.4 that

THEOREM 2.12. *The results in Theorems 2.10 and 2.11 are still true if the range condition (A1) is replaced by the weaker range condition (A1)′ below, provided that the initial condition* $x \in \overline{D(A)}$ *is changed to the condition* $x \in D(A)$, *and that the condition (F1) is changed to the condition (F2). Here*

(A1)′ *For each samll* $0 < \lambda < \lambda_0$, *where* λ_0 *is some positive number satisfying* $\lambda_0 \omega < 1$, *the range of* $(I - \lambda A)$ *contains* $D(A)$.

3. Examples

Three examples will be considered. The first one is about a linear, nonhomogeneous, parabolic, initial-boundary value problem of space dimension one, and the second one is about its analogue of higher space dimensions. The last example concerns a nonlinear, nonhomogeneous, parabolic, initial-boundary value problem of space dimension one. More complex examples will be a subject of other chapters.

EXAMPLE 3.1. Solve for $u = u(x,t)$:

$$u_t(x,t) = u_{xx}(x,t) + f_0(x),$$
$$(x,t) \in (0,1) \times (0,\infty);$$
$$u_x(0,t) = \beta_0 u(0,t), \quad u_x(1,t) = -\beta_1 u(1,t); \tag{3.1}$$
$$u(x,0) = u_0(x);$$

where β_0 and β_1 are two positive constants, and

$$u_t(x,t) \equiv \frac{\partial}{\partial t}u, \quad u_x(x,t) \equiv \frac{\partial}{\partial x}u;$$
$$u_{xx}(x,t) \equiv \frac{\partial^2}{\partial x^2}u.$$

Solution. Define the linear operator

$$F : D(F) \subset C[0,1] \longrightarrow C[0,1]$$

by $Fv = v''$ for

$$v \in D(F)$$
$$\equiv \{w \in C^2[0,1] : w'(j) = (-1)^j \beta_j w(j), j = 0,1\}.$$

It will be shown that F is a closed operator satisfying both the dissipativity condition (B2) and the range condition (B1). As a result, the following two cases are true:

Case 1: $f_0(x) \equiv 0$. [25] In this case, we have, from Theorem 2.1, that (3.1) has a unique solution given by

$$u(t) = \lim_{n\to\infty} (I - \frac{t}{n}F)^{-n} u_0,$$

if $u_0 \in D(F)$ satisfies $Fu_0 = u_0'' \in \overline{D(F)}$. More smoothness of $u(t)$ in t follows from Theorem 2.2, if we further restrict u_0.

Case 2: otherwise. In this case, similar results are true if $f_0(x)$ satisfies the requirements in Theorems 2.7 and 2.8. For example, if $f_0, Fu_0 \in D(F)$, then, by Theorem 2.7, the equation (3.1) has a unique classical solution

$$v(t) = \lim_{n\to\infty} (I - \frac{t}{n}\tilde{F})^{-n} u_0.$$

Here $\tilde{F}w = Fw + f_0$ for $w \in D(\tilde{F}) = D(F)$.

We now begin the proof, which is composed of three steps.

Step 1. (F satisfies the dissipativity condition (B2).) Let v_1 and v_2 be in $D(F)$, and let $v_1 \neq v_2$ to avoid triviality. By the first and second derivative tests, there result, for some $x_0 \in (0,1)$,

$$\|v_1 - v_2\|_\infty = |(v_1 - v_2)(x_0)|;$$

$$(v_1 - v_2)'(x_0) = 0;$$
$$(v_1 - v_2)(x_0)(v_1 - v_2)''(x_0) \le 0.$$

Here $x_0 \in \{0, 1\}$ is impossible, due to the boundary conditions in $D(F)$. For, if $x_0 = 0$ and $\|v_1 - v_2\|_\infty = (v_1 - v_2)(0)$, then

$$(v_1 - v_2)'(0) = \beta_0(v_1 - v_2)(0) > 0,$$

so $(v_1 - v_2)(0)$ cannot be the positive maximum. This contradicts

$$(v_1 - v_2)(0) = \|v_1 - v_2\|_\infty.$$

Other cases can be treated similarly.

The dissipativity condition (B2) is then satisfied, as the calculations show:

$$(v_1 - v_2)(x_0)(Fv_1 - Fv_2)(x_0) \le 0;$$
$$\|v_1 - v_2\|_\infty^2$$
$$= (v_1 - v_2)(x_0)(v_1 - v_2)(x_0)$$
$$\le [(v_1 - v_2)(x_0)]^2 - \lambda(v_1 - v_2)(x_0)(Fv_1 - Fv_2)(x_0)$$
$$\le \|v_1 - v_2\|_\infty\|(v_1 - v_2) - \lambda(Fv_1 - Fv_2)\|_\infty$$
$$\text{for all } \lambda > 0.$$

Step 2. From the theory of ordinary differential equations [5], [24, Corollary 2.13, Chapter 4], the range of $(I - \lambda F), \lambda > 0$, equals $C[0, 1]$, so F satisfies the range condition (B1).

Step 3. (F is a closed operator.) Let $v_n \in D(F)$ converge to v, and Fv_n to w. Then, there is a positive constant K, such that

$$\|v_n\|_\infty \le K;$$
$$\|Fv_n\|_\infty = \|v_n''\|_\infty \le K.$$

This, together with the interpolation inequality [1], [13, page 135], implies that $\|v_n'\|_\infty$ and then $\|v_n\|_{C^2[0,1]}$ are uniformly bounded. Hence it follows from the Ascoli-Arzela theorem [33] that a subsequence of v_n' and then itself converge to v'. That $v \in D(F)$ and $Fv_n = v_n''$ converges to $v'' = Fv$ will be true, whence F is closed. For, by uniform covergence theorem [2],

$$v_n'(x) = \int_{y=0}^{x} v_n''(y)\, dy + v_n'(0)$$

converges to

$$v'(x) = \int_{y=0}^{x} w(y)\, dy + v'(0),$$

and

$$v_n'(j) = (-1)^j \beta_j v_n(j)$$

converges to

$$v'(j) = (-1)^j \beta_j v(j), \quad j = 0, 1;$$

so

$$v \in D(F) \text{ and } Fv = v'' = w$$

by the fundamental theorem of calculus [2].

The proof is complete. $\square$

EXAMPLE 3.2. Solve for $u = u(x,t)$:

$$u_t(x,t) = \triangle u(x,t) + f_0(x),$$
$$(x,t) \in \Omega \times (0,\infty);$$
$$\frac{\partial}{\partial \hat{n}} u(x,t) + \beta_2 u(x,t) = 0, \quad x \in \partial\Omega;$$
$$u(x,0) = u_0(x);$$

(3.2)

where Ω is a bounded, smooth domain in $\mathbb{R}^N$, and $N \geq 2$ is a positive integer; $x = (x_1, x_2, \ldots, x_N)$, $\triangle u = \sum_{i=1}^{N} \frac{\partial^2}{\partial x_i} u$, and $u_t = \frac{\partial}{\partial t} u$; $\partial\Omega$ is the boundary of Ω, and $\frac{\partial}{\partial \hat{n}} u$ is the outer normal derivative of u; β_2 is a positive number.

Solution. Define the linear operator

$$G : D(G) \subset C(\overline{\Omega}) \longrightarrow C(\overline{\Omega})$$

by $Gv = \triangle v$ for

$$v \in D(G)$$
$$\equiv \{w \in C^{2+\mu}(\overline{\Omega}) : \frac{\partial w}{\partial \hat{n}} + \beta_2 w = 0 \quad \text{for } x \in \partial\Omega\}.$$

Here $0 < \mu < 1$, is a constant.

It will be shown that G satisfies both the dissipativity condition (B2) and the weaker range condition $(B1)'$. Consequently, the following two cases are true:

Case 1: $f_0 \equiv 0$. [**25**] In this case, we have, from Theorem 2.3, that the quantity

$$u(t) = \lim_{n\to\infty} (I - \frac{t}{n}G)^{-n} u_0$$

exists if $u_0 \in D(G)$. If $u_0 \in D(G^2)$, further estimates will be derived in order for $u(t)$ to the unique solution of (3.2). More smoothness of $u(t)$ in t will then follow from Theorem 2.3 again, if we impose more restrictions on u_0.

Case 2: otherwise. In this case, it will be shown that if $f_0 \in C^\mu(\overline{\Omega})$, then f_0 satisfies the weaker condition $(F0)'$. Thus, Theorem 2.9 assures the existence of the limit

$$u(t) = \lim_{n\to\infty} (I - \frac{t}{n}\tilde{G})^{-n} u_0$$
$$= \lim_{\nu\to 0} (I - \nu\tilde{G})^{-[\frac{t}{\nu}]} u_0$$

for $u_0 \in D(G)$, where the corresponding nonlinear operator $\tilde{G}$ satisfies the dissipativity condition (A2) and the weaker range condition $(A1)'$, defined by $\tilde{G}v = Gv + f_0$ for $v \in D(\tilde{G}) \equiv D(G)$. Further estimates will be derived under additional assumptions on u_0 and f_0, so that the $u(t)$ is in fact a unique classical solution.

We now begin the proof, which consists of eight steps.

Step 1. (G satisfies the dissipativity condition (B2).) Let v_1 and v_2 be in $D(G)$, and let $v_1 \neq v_2$ to avoid triviality. By the first and second derivative tests, there result, for some $x_0 \in \Omega$,

$$\|v_1 - v_2\|_\infty = |(v_1 - v_2)(x_0)|;$$
$$\nabla(v_1 - v_2)(x_0) = 0, \quad \text{(the gradient of } (v_1 - v_2));$$

$$(v_1 - v_2)(x_0)\triangle(v_1 - v_2)(x_0) \leq 0.$$

Here $x_0 \in \partial\Omega$ is impossible, due to the boundary condition in $D(G)$. For, if $x_0 \in \partial\Omega$ and $\|v_1 - v_2\|_\infty = (v_1 - v_2)(x_0)$, then

$$\frac{\partial}{\partial\hat{n}}(v_1 - v_2)(x_0) > 0$$

by the Hopf boundary point lemma [13]. But this is a contradiction to

$$\frac{\partial}{\partial\hat{n}}(v_1 - v_2)(x_0) = -\beta_2(v_1 - v_2)(x_0) < 0.$$

The case where $x_0 \in \partial\Omega$ and $\|v_1 - v_2\|_\infty = -(v_1 - v_2)(x_0)$ is similar.

The dissipativity condition (B2) is then satisfied, as the calculations show:

$$(v_1 - v_2)(x_0)(Gv_1 - Gv_2)(x_0) \leq 0;$$

$$\|v_1 - v_2\|_\infty^2$$
$$= (v_1 - v_2)(x_0)(v_1 - v_2)(x_0)$$
$$\leq [(v_1 - v_2)(x_0)]^2 - \lambda(v_1 - v_2)(x_0)(Gv_1 - Gv_2)(x_0)$$
$$\leq \|v_1 - v_2\|_\infty \|(v_1 - v_2) - \lambda(Gv_1 - Gv_2)\|_\infty$$
$$\text{for all } \lambda > 0.$$

Step 2. From the theory of linear, elliptic partial differential equations [13], the range of $(I - \lambda G), \lambda > 0$, equals $C^\mu(\overline{\Omega})$, so G satisfies the weaker range condition (B1)$'$ on account of $C^\mu(\overline{\Omega}) \supset D(G)$.

Step 3. It will be shown that $\|u_i\|_{C^{3+\eta}(\overline{\Omega})}, 0 < \eta < 1$, is uniformly bounded if $u_0 \in D(G^2)$, where

$$u_i = (I - \nu G)^{-i}u_0$$

is that in the discretized equation (2.1) in which A is replaced by G.

Let $u_0 \in D(G)$ for a moment. By the dissipativity condition (B2) or using Lemma 4.2 in Section 4, we have

$$\|Gu_i\|_\infty = \|\frac{u_i - u_{i-1}}{\nu}\|_\infty \leq \|Gu_0\|_\infty, \qquad (3.3)$$

which, together with relation

$$u_i - u_0 = \sum_{j=1}^{i}(u_j - u_{j-1}),$$

yields a uniform bound for $\|u_i\|_\infty$. Hence, a uniform bound exists for $\|u_i\|_{C^{1+\lambda}(\overline{\Omega})}$ for any $0 < \lambda < 1$, on using the proof of (4.1) in Chapter 5. (Alternatively, it follows that $\|u_i\|_{W^{2,p}(\Omega)}$ is uniformly bounded for any $p > 2$, on using the L^p elliptic estimates [37]. Hence, so is

$$\|u_i\|_{C^{1+\eta}(\overline{\Omega})} = \|(I - \nu G)^{-i}u_0\|_{C^{1+\eta}(\overline{\Omega})},$$
$$0 < \eta < 1, \qquad (3.4)$$

as a result of the Sobolev embedding theorem [1, 13].) This, applied to the relation

$$Gu_i = (I - \nu G)^{-i}(Gu_0), \qquad (3.5)$$

shows the same thing for $\|Gu_i\|_{C^{1+\eta}(\overline{\Omega})}$, if $Gu_0 \in D(G)$, that is, if $u_0 \in D(G^2)$. Therefore, $\|u_i\|_{C^{3+\eta}(\overline{\Omega})}$ is uniformly bounded if $u_0 \in D(G^2)$, on employing the Schauder global regularity theorem [13, page 111].

Step 4. (Existence of a solution) The result in Step 3, together with the Ascoli-Arzela theorem [**33**], implies that, on putting $i = [\frac{t}{\nu}]$, a subsequence of u_i and then itself, converge to $u(t)$ as $\nu \longrightarrow 0$, with respect to the topology in $C^{3+\lambda}(\overline{\Omega})$ for any $0 < \lambda < 1$. Consequently, as in (6.4), (6.5), and (6.6) in Section 6, we have eventually

$$\frac{du(t)}{dt} = Bu(t) = \lim_{n\to\infty} (I - \frac{t}{n}G)^{-n}(Gu_0);$$
$$u(0) = u_0.$$

Thus $u(t)$ is a solution.

Step 5. (Uniqueness of a solution) Let $v(t)$ be another solution. Then, by the first and second derivative tests, we have, for $x_0 \in \Omega$,

$$\|u(t) - v(t)\|_\infty = |[u(t) - v(t)](x_0)|;$$
$$\nabla[u(t) - v(t)](x_0) = 0;$$
$$[u(t) - v(t)](x_0)\triangle[u(t) - v(t)](x_0) \le 0.$$

Thus it follows that

$$\frac{d}{dt}\|u(t) - v(t)\|_\infty^2$$
$$= \frac{d}{dt}[u(t) - v(t)]^2(x_0)$$
$$= 2[u(t) - v(t)](x_0)\frac{d}{dt}[u(t) - v(t)](x_0)$$
$$= 2[u(t) - v(t)](x_0)[Gu(t) - Gv(t)](x_0)$$
$$\le 0.$$

This implies

$$\|u(t) - v(t)\|_\infty \le \|u(0) - v(0)\|_\infty$$
$$= \|u_0 - u_0\|_\infty = 0,$$

from which uniqueness of a solution results.

Step 6. If f_0 is in $C^\mu(\overline{\Omega})$, then f_0 satisfies $(F0)'$. This is because $(y + \lambda f_0)$ is in $C^\mu(\overline{\Omega})$ for $y \in D(G)$, but $C^\mu(\overline{\Omega})$ is contained in the range of $(I - \lambda G), \lambda > 0$, by Step 2.

Step 7. (Further estimates under additional assumptions on $f_0(x)$ and u_0) We assume additionally that f_0 is in $C^{2+\mu}(\overline{\Omega})$, and that u_0 is in $D(\tilde{G})$ with

$$\tilde{G}u_0 = \triangle u_0 + f_0 \in D(\tilde{G}).$$

Let, for $v_0 \equiv u_0 \in D(G)$,

$$v_i = (I - \nu\tilde{G})^{-i}u_0$$

be that in the discretized equation (2.1), in which A is replaced by $\tilde{G}$ and u_i is replaced by v_i. For convenience, we also define

$$v_{-1} = (I - \nu\tilde{G})v_0.$$

Here $\tilde{G}$ is readily seen to satisfy the weaker range condition $(A1)'$ and the dissipativity condition (A2). Hence, as in Step 3, applying (A2) or using Lemma 4.2 in

Section 4 results in a uniform bound for

$$\|\tilde{G}v_i\|_\infty = \|\frac{v_i - v_{i-1}}{\nu}\|_\infty$$

$$\text{and } \|v_i\|_{C^{1+\eta}(\overline{\Omega})};$$

$$0 < \eta < 1, \quad i = 1, 2, \ldots.$$

On the other hand, because of $\tilde{G}v_0 = \tilde{G}u_0 \in D(\tilde{G})$, v_i satisfies

$$v_i - \nu[\triangle v_i + f_0(x)] = v_{i-1}, \quad x \in \Omega,$$

$$i = 0, 1, \ldots;$$

$$\frac{\partial v_i}{\partial \hat{n}} + \beta_2 v_i = 0, \quad x \in \partial\Omega,$$

$$i = -1, 0, 1, \ldots.$$

Hence, it follows, on letting $w_i = \triangle v_i$ for $i = 0, 1, \ldots$, that

$$w_i - \lambda[\triangle w_i + \triangle f_0(x)] = w_{i-1}, \quad x \in \Omega,$$

$$i = 1, 2, \ldots;$$

$$\frac{\partial w_i}{\partial \hat{n}} + [\frac{\partial f_0}{\partial \hat{n}} + \beta_2 f_0] = -\beta_2 w_i, \quad x \in \partial\Omega,$$

$$i = 0, 1, \ldots.$$

Here

$$w_i - f_0(x) = \frac{v_i - v_{i-1}}{\lambda}, \quad i = 0, 1, \ldots.$$

This induces a nonlinear, dissipative operator

$$\tilde{\tilde{G}} : D(\tilde{\tilde{G}}) \subset C(\overline{\Omega}) \longrightarrow C(\overline{\Omega}),$$

defined by $\tilde{\tilde{G}}v = \triangle v + \triangle f_0(x)$ for

$$v \in D(\tilde{\tilde{G}})$$

$$\equiv \{w \in C^{2+\mu}(\overline{\Omega}) : \frac{\partial w}{\partial \hat{n}} + [\frac{\partial f_0}{\partial \hat{n}} + \beta_2 f_0]$$

$$= -\beta_2 w, \quad x \in \partial\Omega\}.$$

Thus, as in Step 3 again, applying dissipativity condition (A2) or using Lemma 4.2 in Section 4 yields a uniform bound for

$$\|\triangle w_i + \triangle f_0(x)\|_\infty = \|\tilde{\tilde{G}}w_i\|_\infty$$

$$= \|\frac{w_i - w_{i-1}}{\lambda}\|_\infty;$$

$$\|w_i\|_{C^{1+\eta}(\overline{\Omega})}, \quad 0 < \eta < 1, \quad i = 0, 1, \ldots,$$

from which so is yielded for

$$\|v_i\|_{C^{3+\eta}(\overline{\Omega})}, \quad i = 0, 1, \ldots$$

by the Schauder global regularity theorem [**13**, page 111].

   Step 8. That, on putting $i = [\frac{t}{\nu}]$, $u(t) = \lim_{\nu \to 0} v_i$ is a unique calssical solution follows as in Steps 4, and 5.

   The proof is complete. $\square$

EXAMPLE 3.3. Solve for $u = u(x,t)$:

$$u_t(x,t) = u_{xx}(x,t) + f_1(x),$$
$$(x,t) \in (0,1) \times (0,\infty);$$
$$u_x(0,t) \in (-1)^j \beta_j(u(j,t)), \quad j = 0,1;$$
$$u(x,0) = u_0(x);$$

(3.6)

where β_0 and β_1 are maximal monotone graphs in $\mathbb{R} \times \mathbb{R}$. This is the nonlinear analogue of the problem in Example 3.1. Here a monotone graph β is a subset of $\mathbb{R} \times \mathbb{R}$ that satisfies

$$(y_2 - y_1)(x_2 - x_1) \geq 0 \quad \text{for } y_i \in \beta(x_i), \quad i = 1,2.$$

This β is a maximal monotone graph, if it is not properly contained in any other monotone graph. In this case, for any $\lambda > 0$,

$$(I + \lambda\beta)^{-1} : \mathbb{R} \longrightarrow \mathbb{R}$$

is single-valued and non-expansive, as readily checked [3]. Here non-expansiveness means

$$|(I + \lambda\beta)^{-1}x - (I + \lambda\beta)^{-1}y| \leq |x - y|$$

for $x,y \in \mathbb{R}$.

Solution. Define the nonlinear operator

$$H : D(H) \subset C[0,1] \longrightarrow C[0,1]$$

by $Hv = v''$ for

$$v \in D(H)$$
$$\equiv \{w \in C^2[0,1] : w'(j) \in (-1)^j \beta_j(w(j)), \quad j = 0,1\}.$$

It will be shown that H satisfies both the dissipativity condition (A2) and the range condition (A1). In consequence, the following two cases are true:

Cases 1: $f_1(x) \equiv 0$. In this case, it follows from Theorem 2.4 that the quantity

$$u(t) = \lim_{n\to\infty} (I - \frac{t}{n}H)^{-n}u_0$$
$$= \lim_{\mu\to 0}(I - \mu H)^{-[\frac{t}{\mu}]}u_0$$

exists if $u_0 \in \overline{D(H)}$. H will be further shown to satisfy the embedding condition (A3) of embeddedly quasi-demi-closedness, so, by Theorem 2.5, $u(t)$ for $u_0 \in D(H)$ is not only a limit solution but also a strong solution of (3.3). This $u(t)$ will also satisfy the middle equation in (3.6).

Case 2: otherwise. In this case, similar results are true if $f_1(x)$ satisfies the requirements in Theorems 2.10 and 2.11. For instance, if $u_0 \in D(H)$ and $f_1 \in C[0,1]$, then f_1 satisfies the condition (F1). So, by Theorem 2.11, the equation (3.3) has a strong solution

$$v(t) = \lim_{n\to\infty} (I - \frac{t}{n}\tilde{H})^{-n}u_0.$$

Here $\tilde{H}w = Hw + f_1$ for $w \in D(\tilde{H}) = D(H)$.

We now begin the proof, which is comprised of four steps.

Step 1 (H satisfies the dissipativity condition (A2).) Let v_1 and v_2 be in $D(H)$, and let $v_1 \neq v_2$ to avoid triviality. By the first and second derivative tests, there result, for some $x_0 \in [0, 1]$,

$$\|v_1 - v_2\|_\infty = |(v_1 - v_2)(x_0)|;$$
$$(v_1 - v_2)'(x_0) = 0;$$
$$(v_1 - v_2)(x_0)(v_1 - v_2)''(x_0) \leq 0.$$

Here $x_0 \in \{0, 1\}$ is possible, due to the boundary conditions in $D(H)$. For, if $x_0 = 0$ and $\|v_1 - v_2\|_\infty = (v_1 - v_2)(0)$, then the monotonicity of β_0 and the positivity of $(v_1 - v_2)(0)$ implies

$$(v_1 - v_2)'(0) \geq 0.$$

From this, there must $(v_1 - v_2)'(0) = 0$ because if $(v_1 - v_2)'(0) > $ occurs, then $(v_1 - v_2)(0)$ cannot be the positive maximum. Other cases can be treated similarly.

The dissipativity condition (A2) is then satisfied, as the calculations show:

$$(v_1 - v_2)(x_0)(Hv_1 - Hv_2)(x_0) \leq 0;$$
$$\|v_1 - v_2\|_\infty^2$$
$$= (v_1 - v_2)(x_0)(v_1 - v_2)(x_0)$$
$$\leq [(v_1 - v_2)(x_0)]^2 - \lambda(v_1 - v_2)(x_0)(Hv_1 - Hv_2)(x_0)$$
$$\leq \|v_1 - v_2\|_\infty \|(v_1 - v_2) - \lambda(Hv_1 - Hv_2)\|_\infty$$
$$\text{for all } \lambda > 0.$$

Step 2. (H satisfies the range condition (A1).) [**35, 15**] Let $h \in C[0, 1]$ be given, and let $\mu > $ be small such that $0 < \mu < [\log(3)]^{-1}$. It will be shown that the equation

$$u - \mu^2 u'' = h \tag{3.7}$$

has a solution $u \in D(H)$.

From the theory of ordinary differential equations [**5, 24**], the equation (3.7) has the general solution

$$u = ae^{\frac{1}{\mu}x} + be^{-\frac{1}{\mu}x} + u_p;$$
$$u_p = -\frac{1}{\mu} \int_{y=0}^{x} \sinh[\frac{1}{\mu}(x - y)]h(y)\, dy,$$

where a and b are two arbitrary constants, and the particular solution u_p is obtained from the variation of constants formula [**5, 24**]. We will choose suitable a and b, so that the corresponding u lies in $D(H)$.

The boundary conditions $u'(0) \in \beta_0(u(0))$ and $u'(1) \in -\beta_1(u(1))$ in $D(H)$, together with the existence of

$$(I + \mu\beta_j)^{-1}, j = 0, 1 : \mathbb{R} \longrightarrow \mathbb{R},$$

require a and b to satisfy

$$b = (I + \mu\beta_0)^{-1}(2a) - a;$$
$$(I + \mu\beta_1)^{-1}\{2be^{-\frac{1}{\mu}}$$
$$+ \frac{1}{\mu}\int_{y=0}^{1}[\cosh(\frac{1-y}{\mu}) - \sinh(\frac{-(1-y)}{\mu})]h(y)\, dy\}$$

$$= ae^{\frac{1}{\mu}} + be^{-\frac{1}{\mu}} - \frac{1}{\mu} \int_{y=0}^{1} \sinh(\frac{1-y}{\mu}) h(y) \, dy.$$

It follows that a and b meet the requirement if a is a fixed point of the nonlinear map

$$T : \mathbb{R} \longrightarrow \mathbb{R},$$

where, for $x \in \mathbb{R}$,

$$Tx = -e^{-\frac{2}{\mu}}[(I + \mu\beta_0)^{-1}(2x) - x]$$
$$+ \frac{1}{\mu} e^{-\frac{1}{\mu}} \int_{y=0}^{1} \sinh(\frac{1-y}{\mu}) h(y) \, dy$$
$$+ e^{-\frac{1}{\mu}} (I + \mu\beta_1)^{-1} \{ 2e^{-\frac{1}{\mu}}[(I + \mu\beta_0)^{-1}(2x) - x]$$
$$+ \frac{1}{\mu} \int_{y=0}^{1} [\cosh(\frac{1-y}{\mu}) - \sinh(\frac{1-y}{\mu})] h(y) \, dy \}.$$

This T has a unique fixed point by the Banach fixed point theorem, because, for $x_1, x_2 \in \mathbb{R}$ and $0 < \mu < [\log(3)]^{-1}$, T is a strict contraction:

$$|Tx_2 - Tx_1| \leq 9e^{-\frac{2}{\mu}} |x_2 - x_1|$$
$$< |x_2 - x_1|.$$

Here the non-expansiveness of

$$(I + \mu\beta_j)^{-1}, \quad j = 0, 1 : \mathbb{R} \longrightarrow \mathbb{R}$$

was used.

Step 3. (*H* satisfies the embedding condition (A3) of embeddedly quasi-demi-closedness.) [20] Let $v_n \in D(H)$ converge to v in $C[0,1]$, and let $\|Hv_n\|_\infty$ be uniformly bounded. It will be shown that, for each η in the self-dual space $L^2(0,1) = (L^2(0,1))^*$, $\eta(Hv)$ exists and

$$|\eta(Hv_n) - \eta(Hv)| \longrightarrow 0.$$

Here $(C[0,1]; \|\cdot\|_\infty)$ is continuously embedded into $L^2(0,1); \|\cdot\|)$.

Since $\|v_n\|_\infty$ and $\|Hv_n\|_\infty$ are uniformly bounded, so is $\|v_n\|_{C^2[0,1]}$ by the interpolation inequality [1], [13, page 135]. Hence, by the Ascoli-Arzela theorem [33], a subsequence of v_n and then itself converge in $C^1[0,1]$ to v. Also, v_n is uniformly bounded in the Hilbert space $W^{2,2}(0,1)$, whence, by Alaoglu theorem [36], a subsequence of v_n and then itself converge weakly to v [36]. It follows that, for each $\eta \in L^2(0,1)$,

$$|\eta(Hv_n) - \eta(Hv)| = |\int_0^1 (v_n'' - v'')\eta \, dx|$$
$$\longrightarrow 0.$$

Therefore H satisfies the embedding condition (A3).

Step 4. ($u(t)$ satisfies the middle equation in (3.6).) Consider the discretized equation

$$u_i - \nu H u_i = u_{i-1},$$
$$u_i \in D(H),$$

where $i = 1, 2, \ldots,$ $\nu > 0$ satisfies $\nu\omega < 1$, and

$$u_i = (I - \nu H)^{-i} u_0$$

exists uniquely by the range condition (A1) and the dissipativity condition (A2) (see Section 4).

On putting $i = [\frac{t}{\nu}]$, it follows that

$$\lim_{\nu \to 0} u_i = \lim_{\nu \to 0} (I - \nu H)^{-[\frac{t}{\nu}]} u_0 = u(t).$$

On the other hand, by utilizing the dissipativity condition (A2), we have

$$\|u_i''\|_\infty = \|H u_i\|_\infty = \|\frac{u_i - u_{i-1}}{\nu}\|_\infty$$
$$\leq \|H u_0\|_\infty.$$

This, combined with the relation

$$u_i - u_0 = \sum_{j=1}^{i}(u_j - u_{j-1}),$$

yields a bound for $\|u_i\|_\infty$. Those, in turn, result in a bound for $\|u_i\|_{C^2[0,1]}$ by the interpolation inequality [1], [13, page 135]. Therefore it follows from Ascoli-Arzela theorem [33] that a subsequence of u_i and then itself converge to a limit in $C^1[0, 1]$, as $\nu \longrightarrow 0$. This limit equals $u(t)$ as shown above. Consequently, $u(t)$ satisfies the middle equation in (3.6), as u_i does so.

The proof is complete. $\square$

4. Some Preliminary Results

Let the nonlinear, multi-valued operator A in Section 1 satisfy the range condition (A1) and the dissipativity (A2).

Let D_μ be the range of $(I - \mu A)$ where $\mu \in \mathbb{R}$. For $x \in D_\mu$, let J_μ be such that $J_\mu x = (I - \mu A)^{-1} x$.

In this section, some properties of J_μ will be explored, with the aid of which a recursive inequality will be established. Then, by using a basic theory of elementary difference equations in Section 5, we solve this inequality in Section 6. Once this inequality is solved, the proof of the main results follows.

LEMMA 4.1. Let $\mu > 0$ be such that $\mu\omega < 1$. Then the function

$$J_\mu = (I - \mu A)^{-1} : D_\mu \longrightarrow D(A)$$

is single-valued, and for $x, y \in D_\mu$, the inequality

$$\|J_\mu x - J_\mu y\| \leq (1 - \mu\omega)^{-1}\|x - y\|$$

is true.

PROOF. Let $x, y \in D_\mu$ and let $v \in J_\mu x, w \in J_\mu x$. Then $v = w$, so J_μ is single-valued. This is because

$$v - \mu A v \ni x \quad \text{and} \quad w - \mu A w \ni x,$$

from which

$$(v - w) - \frac{\mu}{1 - \mu\omega}[(A - \omega)v - (A - \omega)w] \ni 0.$$

By virtue of the dissipativity condition (A2), we have

$$\|v - w\| \leq 0,$$

giving $v = w$.

Similarly, let $u = J_\mu y$, and the desired inequality follows. This is because

$$v - \mu Av \ni x \text{ and } u - \mu Au \ni y,$$

whence

$$(v - u) - \mu(1 - \mu\omega)^{-1}[(A - \omega)v - (A - \omega)u]$$
$$\ni (1 - \mu\omega)^{-1}(x - y).$$

Using the dissipativity condition (A2), it follows that

$$\|v - u\| \leq (1 - \mu\omega)^{-1}\|x - y\|.$$

$\square$

LEMMA 4.2. *Let $\mu > 0$ be such that $\mu\omega < 1$. Then, for $n \in \mathbb{N}$ and $x \in D(A) \cap D_\mu^n$ where D_μ^n is the range of $(I - \mu A)^n$, the inequalities*

$$\|J_\mu x - x\| \leq \mu(1 - \mu\omega)^{-1}|Ax|;$$
$$\|J_\mu^n x - J_\mu^{n-1} x\| \leq \mu(1 - \mu\omega)^{-n}|Ax|;$$

$$\|J_\mu^n x - x\| \leq \begin{cases} n\mu(1 - \mu\omega)^{-n}|Ax|, & \text{if } \omega \geq 0; \\ n\mu|Ax|, & \text{if } \omega \leq 0; \end{cases}$$

are true, where $|Ax| \equiv \inf\{\|y\| : y \in Ax\}$.

PROOF. Let $y \in Ax$. Then

$$x - \mu y = x - \mu y \in D_\mu,$$

so by Lemma 4.1, we have

$$x = J_\mu(x - \mu y),$$
$$\|J_\mu x - x\| = \|J_\mu x - J_\mu(x - \mu y)\|$$
$$\leq (1 - \mu\omega)^{-1}\|y\|.$$

Since $y \in Ax$ is arbitrary, the first desired inequality follows.

The second and third inequalities are proved by repeated use of Lemma 4.1. For,

$$\|J_\mu^n x - J_\mu^{n-1} x\| \leq (1 - \mu\omega)^{-1}\|J_\mu^{n-1} x - J_\mu^{n-2} x\|$$
$$\leq \cdots \leq \mu(1 - \mu\omega)^{-n}|Ax|;$$

and

$$\|J_\mu^n x - x\| \leq \|J_\mu^n x - J_\mu^{n-1} x\| + \|J_\mu^{n-1} x - J_\mu^{n-2} x\| + \cdots$$
$$+ \|J_\mu x - x\|$$
$$\leq (1 - \mu\omega)^{-(n-1)}\|J_\mu x - x\|$$
$$+ (1 - \mu\omega)^{-(n-2)}\|J_\mu x - x\| + \cdots$$
$$+ \|J_\mu x - x\|$$

$$\leq \begin{cases} n\mu(1 - \mu\omega)^{-n}|Ax|, & \text{if } \omega \geq 0; \\ n\mu|Ax|, & \text{if } \omega \leq 0. \end{cases}$$

The proof is complete. $\qquad\square$

LEMMA 4.3 (Nonlinear resolvent identity). *Let* $\lambda, \mu > 0$ *be such that* $\lambda\omega, \mu\omega < 1$. *Then for* $x \in D_\lambda$, *the quantity*

$$\frac{\mu}{\lambda}x + \frac{\lambda - \mu}{\lambda}J_\lambda x$$

lies in D_μ, *and the identity*

$$J_\lambda x = J_\mu(\frac{\mu}{\lambda}x + \frac{\lambda - \mu}{\lambda}J_\lambda x)$$

is true.

PROOF. Let $x \in D_\lambda$, for which there is an $x_0 \in D(A)$ and a $y_0 \in Ax_0$, such that

$$x_0 - \lambda y_0 = x.$$

Hence $x_0 = J_\lambda x$, and

$$\frac{\mu}{\lambda}x + \frac{\lambda - \mu}{\lambda}J_\lambda x = \frac{\mu}{\lambda}(x_0 - \lambda y_0) + \frac{\lambda - \mu}{\lambda}x_0$$
$$= x_0 - \mu y_0$$
$$\in D_\mu.$$

Consequently,

$$J_\lambda x = x_0 = J_\mu(\frac{\mu}{\lambda}x + \frac{\lambda - \mu}{\lambda}J_\lambda x).$$

Here the functions J_λ and J_μ are single-valued by Lemma 4.1. $\qquad\square$

LEMMA 4.4. *Let* $\lambda > \mu > 0$ *be such that* $\lambda\omega, \mu\omega < 1$. *Then, for* $x \in D_\lambda \cap D_\mu$,

$$(1 - \lambda\omega)\|A_\lambda x\| \leq (1 - \mu\omega)\|A_\mu x\|$$

holds. Here A_λ *denotes the operator* $\lambda^{-1}(I - J_\lambda)$.

PROOF. [8] The result follows from this calculation

$$\|A_\lambda x\| = \frac{\|x - J_\lambda x\|}{\lambda} \leq \frac{\|x - J_\mu x\|}{\lambda} + \frac{\|J_\mu x - J_\lambda x\|}{\lambda}$$
$$\leq \frac{\mu}{\lambda}\|A_\mu x\| + \frac{1}{\lambda}\|J_\mu x - J_\mu(\frac{\mu}{\lambda}x + \frac{\lambda - \mu}{\lambda}J_\lambda x)\|$$
$$\leq \frac{\mu}{\lambda}\|A_\mu x\| + \frac{\lambda - \mu}{\lambda}(1 - \mu\omega)^{-1}\|\frac{x - J_\lambda x}{\lambda}\|$$
$$= \frac{\mu}{\lambda}\|A_\mu x\| + \frac{\lambda - \mu}{\lambda}(1 - \mu\omega)^{-1}\|A_\lambda x\|.$$

Here Lemmas 4.1 and 4.3 were used. $\qquad\square$

REMARK 4.5. By the assumption (A1), the range of $(I - \lambda A)$ contains $\overline{D(A)}$ for small $0 < \lambda < \lambda_0$. Thus it follows from Lemma 4.4 that, for $x \in \overline{D(A)}$, the limit

$$\lim_{\lambda \to 0}\|A_\lambda x\| = \lim_{\lambda \to 0}\|\frac{x - J_\lambda x}{\lambda}\|$$

exists and can equal ∞. This will be used in Chapter 2.

LEMMA 4.6. *Let $\lambda \geq \mu > 0$ be such that $\lambda\omega, \mu\omega < 1$. Then for $x \in D_\lambda^m \cap D_\mu^n$ and $n, m \in \mathbb{N}$, the inequality*

$$a_{m,n} \leq \gamma\alpha a_{m-1,n-1} + \gamma\beta a_{m,n-1}$$

is true, where

$$a_{m,n} \equiv \|J_\mu^n x - J_\lambda^m x\|, \gamma \equiv (1 - \mu\omega)^{-1};$$

$$\alpha \equiv \frac{\mu}{\lambda}, \quad \beta \equiv 1 - \alpha.$$

PROOF. Using Lemmas 4.3 and 4.1, it follows that

$$a_{m,n} = \|J_\mu^n - J_\mu(\frac{\mu}{\lambda}J_\lambda^{m-1}x + \frac{\lambda-\mu}{\lambda}J_\lambda^m x)\|$$

$$\leq (1 - \mu\omega)^{-1}\|J_\mu^{n-1}x - (\frac{\mu}{\lambda}J_\lambda^{m-1}x + \frac{\lambda-\mu}{\lambda}J_\lambda^m x)\|$$

$$\leq (1 - \mu\omega)^{-1}\{\frac{\mu}{\lambda}\|J_\mu^{n-1}x - J_\lambda^{m-1}x\| + \frac{\lambda-\mu}{\lambda}\|J_\mu^{n-1}x - J_\lambda^m x\|\}$$

$$= \gamma\alpha a_{m-1,n-1} + \gamma\beta a_{m,n-1}.$$

$\square$

LEMMA 4.7. *Let $\alpha, \beta > 0$ be such that $\alpha + \beta = 1$, and let positive integers n, m satisfy $n \geq m$. Then the inequality*

$$\sum_{j=0}^m \binom{n}{j}\alpha^j\beta^{n-j}(m-j) \leq \sqrt{(n\alpha - m)^2 + n\alpha\beta}$$

is true.

PROOF. The proof will be divided into three steps.
Step 1. Using the Schwartz inequality, we have

$$\sum_{j=0}^m \binom{n}{j}\alpha^j\beta^{n-j}(m-j) \leq \sum_{j=0}^n \binom{n}{j}\alpha^j\beta^{n-j}|m-j|$$

$$\leq \left(\sum_{j=0}^n \binom{n}{j}\alpha^j\beta^{n-j}\right)^{\frac{1}{2}} \left(\sum_{j=0}^n \binom{n}{j}\alpha^j\beta^{n-j}(m-j)^2\right)^{\frac{1}{2}}. \qquad (4.1)$$

Step 2. The relations are true:

$$\sum_{j=0}^n \binom{n}{j}\alpha^j\beta^{n-j} = (\alpha + \beta)^n;$$

$$\sum_{j=0}^n \binom{n}{j}j\alpha^j\beta^{n-j} = \alpha n(\alpha + \beta)^{n-1};$$

$$\sum_{j=0}^n \binom{n}{j}j^2\alpha^j\beta^{n-j} = \alpha^2 n(n-1)(\alpha + \beta)^{n-2} + \alpha n(\alpha + \beta)^{n-1}.$$

The first relation is the binomial theorem, the second follows from the differentiation of the first, with respect to α, and the third is the result of differentiating the second, with respect to α.

Step 3. The relations in Step 2, together with $\alpha + \beta = 1$, applied to the right side of (4.1), complete the proof. $\square$

The material of this section is taken from Crandall-Liggett [**6**, pages 268-271].

5. Difference Equations Theory

We now introduce a basic part of the theory of difference equations [**28**]. Let

$$\{b_n\} = \{b_n\}_{n \in \{0\} \cup \mathbb{N}} = \{b_n\}_{n=0}^{\infty}$$

be a sequence of real numbers. For such a sequence $\{b_n\}$, we further extend it by defining

$$b_n = 0, \text{ if } n = -1, -2, \ldots.$$

The set of all such sequences $\{b_n\}$'s will be denoted by S. Thus, if $\{a_n\} \in S$, then

$$0 = a_{-1} = a_{-2} = \cdots.$$

Define a right shift operator $E : S \longrightarrow S$ by

$$E\{b_n\} = \{b_{n+1}\} \quad \text{for } \{b_n\} \in S.$$

Similarly, define a left shift operator $E^{\#} : S \longrightarrow S$ by

$$E^{\#}\{b_n\} = \{b_{n-1}\} \quad \text{for } \{b_n\} \in S.$$

For $c \in \mathbb{R}$ and $c \neq 0$, define the operator $(E - c)^* : S \longrightarrow S$ by

$$(E - c)^*\{b_n\} = \{c^n \sum_{i=0}^{n-1} \frac{b_i}{c^{i+1}}\}$$

for $\{b_n\} \in S$. Here the first term on the right side of the equality, corresponding to $n = 0$, is zero.

One more definition is that define, for $\{b_n\} \in S$,

$$(E - c)^{i*}\{b_n\} = [(E - c)^*]^i\{b_n\}, \quad i = 1, 2, \ldots;$$
$$E^{i\#}\{b_n\} = (E^{\#})^i\{b_n\}, \quad i = 1, 2, \ldots;$$
$$(E - c)^0\{b_n\} = \{b_n\}.$$

Based on those definitions above, we derive the following results. It will be seen from below that $(E - c)^*$ acts approximately as the inverse of $(E - c)$.

LEMMA 5.1. *Let $\{b_n\}$ and $\{d_n\}$ be in S. Then the following are true:*

$$(E - c)^*(E - c)\{b_n\} = \{b_n - c^n b_0\};$$
$$(E - c)(E - c)^*\{b_n\} = \{b_n\};$$
$$(E - c)^*\{b_n\} \leq (E - c)^*\{d_n\}, \quad \text{if } c > 0 \text{ and } \{b_n\} \leq \{d_n\}.$$

Here $\{b_n\} \leq \{d_n\}$ means $b_n \leq d_n$ for $n = 0, 1, 2, \ldots$.

PROOF. The proof follows from straightforward calculations. $\square$

PROPOSITION 5.2. *Let $\xi, c \in \mathbb{R}$ be such that $c \neq 1$ and $c \neq 0$. Let, be in S, the three sequences $\{n\}_{n=0}^{\infty}, \{c^n\}_{n=0}^{\infty}$, and $\{\xi\}_{n=0}^{\infty}$ of real numbers. Then the following identities are true:*

$$(E - c)^*\{n\} = \{\frac{n}{d} - \frac{1}{d^2} + \frac{c^n}{d^2}\};$$
$$(E - c)^*\{\xi\} = \{\frac{\xi}{d} - \frac{\xi c^n}{d}\};$$

$$(E - c)^{i*}\{c^n\} = \{\binom{n}{i}c^{n-i}\}.$$

Here $d = 1 - c$ and $i = 0, 1, 2, \ldots$.

PROOF. Since, by definition,

$$(E - c)^*\{n\} = \{c^n \sum_{j=0}^{n-1} \frac{j}{c^{j+1}}\}$$

$$= \{c^{n-1}[c^{-1} + 2c^{-2} + \cdots + (n-1)c^{-(n-1)}]\}$$

$$= \{c^{n-1}(-1)c\frac{d}{dc}[c^{-1} + c^{-2} + \cdots + c^{-(n-1)}]\},$$

the first identity follows, using the formula for a finite geometric series.
Because

$$(E - c)^*\{\xi\} = \{\xi c^n(c^{-1} + c^{-2} + \cdots + c^{-n})\},$$

the second identity follows.

Finally, we use mathematical induction to prove the third identity. This identity is true for $i = 0, 1$, due to the calculations

$$(E - c)^0\{c^n\} = \{c^n\} = \{c^{n-0}\binom{n}{0}\};$$

$$(E - c)^*\{c^n\} = \{c^n \sum_{j=0}^{n-1} \frac{c^j}{c^{j+1}}\} = \{nc^{n-1}\}$$

$$= \{\binom{n}{1}c^{n-1}\}.$$

Hence, by assuming it holds for $i = k < n$, we shall show that it continues to hold for $i = k + 1 < n$. The calculations

$$(E - c)^{(k+1)*}\{c^n\} = (E - c)^*(E - c)^{k*}\{c^n\}$$

$$= (E - c)^*\{\binom{n}{k}c^{n-k}\}$$

$$= \{c^n \sum_{j=0}^{n-1} \frac{c^{j-k}\binom{j}{k}}{c^{j+1}}\}$$

$$= \{c^{n-k-1} \sum_{j=0}^{n-1} \binom{j}{k}\},$$

together with the standard combinatorics identity [**4**, page 79] or [**27**, page 52]

$$\binom{r}{r} + \binom{r+1}{r} + \cdots + \binom{n}{r} = \binom{n+1}{r+1}$$

for $r, n \in \mathbb{N}$ and $n \geq r$, imply that the third identity holds for $i = k + 1$.   □

PROPOSITION 5.3. *Let $\xi, c \in \mathbb{R}$ be such that $c \neq 1$ and $c\xi \neq 0$. Let, be in S, the three sequences $\{n\xi^n\}_{n=0}^{\infty}, \{\xi^n\}_{n=0}^{\infty}$, and $\{(c\xi)^n\}_{n=0}^{\infty}$ of real numbers. Then the identities are true:*

$$(E - c\xi)^*\{n\xi^n\} = \{(\frac{n\xi^n}{d} - \frac{\xi^n}{d^2} + \frac{c^n\xi^n}{d^2})\frac{1}{\xi}\};$$

$$(E - c\xi)^*\{\xi^n\} = \{(\frac{\xi^n}{d} - \frac{c^n\xi^n}{d})\frac{1}{\xi}\};$$

$$(E - c\xi)^{i*}\{(c\xi)^n\} = \{\binom{n}{i}(c\xi)^{n-i}\}.$$

Here $d = 1 - c$ and $i = 0, 1, 2, \ldots$.

PROOF. Observe that the third identity in the Proposition 5.3 is the same as that in the Proposition 5.2, so no proof is needed for it.

Thanks to the calculations

$$(E - c\xi)^*\{n\xi^n\} = \{(c\xi)^n \sum_{j=0}^{n-1} \frac{j\xi^j}{(c\xi)^{j+1}}\}$$

$$= \{\xi^{n-1}(E - c)^*n\};$$

$$(E - c\xi)^*\{\xi^n\} = \{(c\xi)^n \sum_{j=0}^{n-1} \frac{\xi^j}{(c\xi)^{j+1}}\}$$

$$= \{\xi^{n-1}(E - c)^*(1)\},$$

the first and second identities follow from applying Proposition 5.2. ☐

The material of this section is taken from our article [**21**].

6. Proof of the Main Results

Within this section, it suffices to consider $\omega \geq 0$, in view of Lemma 4.2. This will be seen from the following proofs.

The nonlinear homogeneous case will be proved first, because, once this is done, the remaining cases become clearer.

6.1. The Nonlinear Homogeneous Case. We shall use the notions of difference equations introduced in Section 5. For a doubly indexed sequence $\{\rho_{m,n}\} = \{\rho_{m,n}\}_{m,n=0}^{\infty} \in S \times S$ of real numbers, define

$$E_1\{\rho_{m,n}\} = \{\rho_{m+1,n}\};$$

$$E_2\{\rho_{m,n}\} = \{\rho_{m,n+1}\}.$$

Thus, E_1 is the right shift operator acting on the first index m, and E_2 is the right shift operator acting on the second index n. It is readily seen that E_1 and E_2 commute:

$$E_1 E_2\{\rho_{m,n}\} = E_2 E_1\{\rho_{m,n}\}.$$

Theorems 2.4 and 2.5 will be proved after the following Lemma 6.1 and Proposition 6.2.

LEMMA 6.1. *Under the assumptions in Lemma 4.6, the inequalities are true:*

$$\{a_{m,n}\} \leq (\alpha\gamma(E_2 - \beta\gamma)^*)^m\{a_{0,n}\}$$

$$+ \sum_{i=0}^{m-1} (\gamma\alpha(E_2 - \gamma\beta)^*)^i\{(\gamma\beta)^n a_{m-i,0}\} \quad \text{for } n \geq m;$$

$$\{a_{m,n}\} \leq (\gamma\beta + \gamma\alpha E_1^{\#})^n\{a_{m,0}\}$$

$$= \gamma^n \sum_{i=0}^{n} \binom{n}{i} \beta^{n-i} \alpha^i E_1^{i\#}\{a_{m,0}\}$$

$$= \{\gamma^n \sum_{i=0}^{n} \binom{n}{i} \beta^{n-i} \alpha^i a_{m-i,0}\} \quad \text{for } m \geq n.$$

PROOF. From Lemma 4.6, we have

$$E_1(E_2 - \beta\gamma)\{a_{m,n}\} \leq \{\alpha\gamma a_{m,n}\},$$

so applying $(E_2 - \beta\gamma)^*$ to both sides of the above and using Lemma 5.1, we readily derive

$$\{a_{m+1,n}\} = E_1\{a_{m,n}\}$$
$$\leq \alpha\gamma(E_2 - \beta\gamma)^*\{a_{m,n}\} + (\gamma\beta)^n E_1\{a_{m,0}\}.$$

This recursive relation will give the first inequality immediately.

On the other hand, if applying $E_1^{\#}$ instead to both sides, we have, on using Lemma 5.1,

$$\{a_{m,n+1}\} = E_2\{a_{m,n}\}$$
$$\leq (\gamma\beta + \gamma\alpha E_1^{\#})\{a_{m,n}\}.$$

This recursive relation will easily deliver the second inequality.   □

PROPOSITION 6.2. *Under the assumptions in Lemma 4.6 with $\lambda_0 > \lambda \geq \mu$, the inequality is true:*

$$a_{m,n} \leq [(n\mu - m\lambda) + \sqrt{(n\mu - m\lambda)^2 + n\mu(\lambda - \mu)}]\gamma^n|Ax|$$
$$+ \gamma^n(1 - \lambda\omega)^{-m}\sqrt{(n\mu - m\lambda)^2 + n\mu(\lambda - \mu)}|Ax|$$
$$\text{for } n, m \geq 0,$$

where $|Ax| = \inf\{\|y\| : y \in Ax\}$.

PROOF. The proof will be divided into two cases.

Case 1: $n \geq m$. By applying Lemmas 4.2, 5.1, and 6.1, and by making repeated use of Proposition 5.3 and carefully grouping terms in which tedious but not difficult calculations are involved, we have, for $n \geq m$,

$$\{a_{0,n}\} \leq \{n\mu\gamma^n|Ax|\};$$

$$((E_2 - \beta\gamma)^*)^m\{n\gamma^n\} = \{\frac{n\gamma^n}{\alpha^m}\frac{1}{\gamma^m} - \frac{m\gamma^n}{\alpha^{m+1}}\frac{1}{\gamma^m}$$

$$+ \left(\sum_{i=0}^{m-1} \binom{n}{i} \frac{\beta^{n-i}}{\alpha^{m+1-i}}(m-i)\frac{1}{\gamma^m}\right)\gamma^n\};$$

$$\sum_{i=0}^{m-1} (\gamma\alpha(E_2 - \gamma\beta)^*)^i\{(\gamma\beta)^n a_{m-i,0}\}$$

$$\leq \{\gamma^n(1 - \lambda\omega)^{-m} \sum_{i=0}^{m-1} \binom{n}{i} \alpha^i \beta^{n-i}(m-i)\lambda|Ax|\}.$$

This, combined with Lemma 4.7, gives the desired result.

Case 2: $m \geq n$. Applying Lemmas 4.2, 5.1, and 6.1, we also have, for $m \geq n$,

$\{a_{m,n}\}$

$$\leq \{\gamma^n (1 - \lambda\omega)^{-m} \sum_{i=0}^{n} \binom{n}{i} \beta^{n-i}\alpha^i[(n-i) + (m-n)]\lambda|Ax|\}$$

$$\leq \{\gamma^n (1 - \lambda\omega)^{-m} (m\lambda - n\mu)|Ax|\}$$

after some calculations.

Cases 1 and 2 complete the proof. $\qquad\square$

PROPOSITION 6.3. *Under the assumptions in Lemma 4.6 with $\lambda_0 > \lambda \geq \mu$, the inequality, if $\omega = 0$, is true for $n, m \geq 0$:*

$$a_{m,n} \leq [(n\mu - m\lambda) + 2\sqrt{(n\mu - m\lambda)^2 + n\mu(\lambda - \mu)}]|Ax|.$$

PROOF. By letting $\omega = 0$ in Proposition 6.2, the result follows. $\qquad\square$

We are now ready for **Proof of Theorem 2.4:**

PROOF. We divide the proof into three steps, where Step 1 consists of two cases.

Step 1. (The existence of $U(t)x$ for $x \in \overline{D(A)}$)

Case 1: $x \in D(A)$. From Proposition 6.2, it follows that

$$\|J_{\frac{t}{n}}^n x - J_{\frac{t}{m}}^m x\|,$$

the $a_{m,n}$ with $\mu = \frac{t}{n} \leq \lambda = \frac{t}{m} < \lambda_0$, converges to zero for finite $t \geq 0$, as $n \geq m \longrightarrow \infty$. Thus

$$J_{\frac{t}{n}}^n x = (I - \frac{t}{n}A)^{-n}x$$

for $x \in D(A)$ is a Cauchy sequence in X, so the limit

$$\lim_{n\to\infty}(I - \frac{t}{n}A)^{-n}x$$

exists for finite $t \geq 0$.

Case 2: $x \in \overline{D(A)}$. Let $x_l \in D(A)$ be such that $x_l \longrightarrow x$ as $l \longrightarrow \infty$. Then, using Lemma 4.1, we have, for $\frac{t}{n} \leq \frac{t}{m} < \lambda_0$,

$$\|J_{\frac{t}{n}}^n x - J_{\frac{t}{m}}^m x\| \leq \|J_{\frac{t}{n}}^n x - J_{\frac{t}{n}}^n x_l\| + \|J_{\frac{t}{n}}^n x_l - J_{\frac{t}{m}}^m x_l\|$$

$$+ \|J_{\frac{t}{m}}^m x_l - J_{\frac{t}{m}}^m x\|$$

$$\leq (1 - \frac{t}{n}\omega)^{-n}\|x - x_l\| + \|J_{\frac{t}{n}}^n x_l - J_{\frac{t}{m}}^m x_l\|$$

$$+ (1 - \frac{t}{m}\omega)^{-m}\|x_l - x\|,$$

and this, together with Case 1, implies, for finite $t \geq 0$,

$$0 \leq \limsup_{n,m\to\infty}\|J_{\frac{t}{n}}^n x - J_{\frac{t}{m}}^m x\|$$

$$\leq 2e^{\omega t}\|x_l - x\| \quad \text{for all } l.$$

Hence, letting $l \longrightarrow \infty$, we deduce

$$\lim_{n,m\to\infty}\|J_{\frac{t}{n}}^n x - J_{\frac{t}{m}}^m x\| = \limsup_{n,m\to\infty}\|J_{\frac{t}{n}}^n x - J_{\frac{t}{m}}^m x\|$$

$$= 0,$$

that is, $(I - \frac{t}{n}A)^{-n}x$ for $x \in \overline{D(A)}$ is a Cauchy sequence in X. Therefore, the limit

$$U(t)x \equiv \lim_{n \to \infty} (I - \frac{t}{n}A)^{-n}x$$

exists for finite $t \geq 0$.

Step 2. On setting $\mu = \lambda = \frac{t}{n}$ and $m = [\frac{t}{\mu}]$ in Proposition 6.2, it follows from the Case 2 above that, for $x \in \overline{D(A)}$,

$$U(t)x = \lim_{\mu \to 0} (I - \mu A)^{-[\frac{t}{\mu}]}x.$$

Here $[a]$ for each $a \in \mathbb{R}$ is the greatest integer that is less than or equal to a.

Step 3. (The continuity and Lipschitz continuity of $U(t)x$, [6, page 272], [30, page 136]) For $x \in \overline{D(A)}$, let $x_l \in D(A)$ be such that $x_l \longrightarrow x$ as $l \longrightarrow \infty$. Then as in the Case 2 above, we have, for $\mu \leq \lambda < \lambda_0$,

$$\|J_\mu^n x - J_\lambda^m x\|$$
$$\leq (1 - \mu\omega)^{-n}\|x - x_l\| + \|J_\mu^n x_l - J_\lambda^m x_l\|$$
$$+ (1 - \lambda\omega)^{-m}\|x_l - x\|.$$

Hence, on setting $n = [\frac{t}{\mu}], m = [\frac{\tau}{\lambda}]$ in the inequality in Proposition 6.2 and letting $\mu, \lambda \longrightarrow \infty$, where $t, \tau \geq 0$, it follows that

$$\|U(t)x - U(\tau)x\|$$
$$\leq (e^{\omega t} + e^{\omega \tau})\|x_l - x\| + |t - \tau|[2e^{\omega t}$$
$$+ e^{\omega(t+\tau)}]|Ax_l|$$

for all l. The continuity of $U(t)x$ in t results, if we let $t \longrightarrow \tau$ first and then $l \longrightarrow \infty$ next. However, if $x \in D(A)$, then the Lipschity continuity of $U(t)x$ is a consequence of setting $x_l = x$ for all l and letting $l \longrightarrow \infty$.

□

Next, **Proof of Theorem 2.5:**

PROOF. We divide the proof into five steps.

Step 1. By Lemma 4.1, the u_i in (2.1) satisfies

$$\|u_i - u_{i-1}\| \leq (1 - \nu\omega)^{-i}\nu\|v_0\|,$$

where $\nu = \frac{T}{n}$ satisfies $\nu < \lambda_0$, $(1 - \nu\omega)^{-i}$ is uniformly bounded by some $K > 0$ for all large $n \in \mathbb{N}$ and $i = 1, 2, \ldots, n$, and $v_0 \in Au_0$ is such that the new element $u_{-1} \equiv u_0 - \nu v_0$ is defined. This is because

$$u_i - u_{i-1} = J_\nu u_{i-1} - J_\nu u_{i-2}.$$

Step 2. (2.3) is an immediate consequence of Step 1 and (2.2).

Step 3. By letting $\mu = \lambda = \frac{t}{n}$ and $m = [\frac{t}{\mu}]$ in Proposition 6.2, it is readily seen from the proof of Theorem 2.4 that

$$\lim_{n \to \infty} (I - \frac{t}{n}A)^{-n}u_0 = \lim_{\mu \to 0} (I - \mu A)^{-[\frac{t}{\mu}]}u_0.$$

Here for each $a \in \mathbb{R}$, $[a]$ is the greatest integer that is less than or equal to a.

Step 4. Claim that $u(t)$ is the uniform limit of $u^n(t)$ on $[0, T]$, where $T > 0$ is arbitrary. For each $t \in [0, T)$, we have $t \in [t_{i-1}, t_i)$ for some i, so $i - 1 = [\frac{t}{\nu}]$. Thus it follows from Step 3 that, for each $t \in [0, T)$,

$$u_{i-1} = (I - \nu A)^{-(i-1)} u_0$$
$$= (I - \nu A)^{-[\frac{t}{\nu}]} u_0$$

converges, as $\nu = \frac{T}{n} \longrightarrow 0$. But the pointwise convergence of u_{i-1} is the same as that of $u^n(t)$, on account of Step 1 and (2.2). Hence, the Ascoli-Arzela theorem [**33**] will prove that $u(t)$ is the uniform limit of $u^n(t)$ on $[0, T]$, as $n \longrightarrow \infty$. This is because Step 1 applied to (2.2) gives

$$\|u^n(t) - u^n(\tau)\| \leq K\|v_0\| |t - \tau|$$

for $t, \tau \in [0, T]$, so $u^n(t)$ is equi-continuous in $C([0, T]; X)$, the real Banach space of continuous functions from $[0, T]$ to X.

Step 5. (Concerning a strong solution) Let $v_n(t) \in A\chi^n(t)$ be such that (2.3) gives

$$\frac{du^n(t)}{dt} = v_n(t)$$

for $t \in (t_{i-1}, t_i]$. Integrating (2.3) yields that, for each $\phi \in Y^* \subset X^*$,

$$\phi(u^n(t) - u_0) = \int_0^t \phi(v^n(\tau)) \, d\tau$$
$$\in \phi\left(\int_0^t A\chi^n(\tau) \, d\tau\right)$$
$$= \int_0^t \phi(A\chi^n(\tau)) \, d\tau,$$

where $\sup_{t \in [0, T]} \|v_n(t)\| \leq K$ by Step 1. Since $u^n(t) \longrightarrow u(t)$ uniformly for bounded t and A is embeddedly quasi-demi-closed, we have that $\phi(v_n(t))$ converges to $\phi(v(t))$ through some subsequence for some $v(t) \in Au(t)$. It then follows from the Lebesgue convergence theorem that

$$\phi(u(t) - u_0) = \int_0^t \phi(v(\tau)) \, d\tau$$
$$= \phi\left(\int_0^t v(\tau) \, d\tau\right),$$

so $u(t) - u_0 = \int_0^t v(\tau) \, d\tau$ in Y. On employing the Radon-Nikodym type theorem [**30**, pages 10-11], [**33**], we see at once that

$$\frac{du(t)}{dt} = v(t) \in Au(t) \quad \text{in } Y$$

$$\text{for almost every } t;$$

$$u(0) = u_0.$$

$\square$

6.2. The Linear Homogeneous Case. Define the operator B as in Section 1, and use the same quantities $J_\mu^n x, J_\lambda^m x, \mu, \lambda, \beta, \alpha$ as in Section 4 but with the operator B replacing A.

Letting $a_{m,n} = J_\mu^n x - J_\lambda^m x$ for $\mu \leq \lambda < \lambda_0$ and for $x \in D(B)$, it follows from the resolvent identity in Lemma 4.3 that

$$a_{m,n} = J_\mu[J_\mu^{n-1}x - (\alpha J_\lambda^{m-1}x + \beta J_\lambda^m x)]$$
$$= J_\mu(\alpha a_{m-1,n-1} + \beta a_{m,n-1}).$$

This is the same as

$$E_1(E_2 - J_\mu\beta)\{a_{m,n}\} = J_\mu\alpha\{a_{m,n}\} \tag{6.1}$$

if used are the right shift operators defined below.

The right shift operators E_1, E_2 and the left shift operator $E_1^\#$ are as in Subsection 6.1 but act on sequences in X. Thus

$$E_1, E_2, E_1^\# : S \times S \longrightarrow S \times S;$$

S is the set of all sequences in X;

$$E_1\{\rho_{m,n}\} = \{\rho_{m+1,n}\}, \quad E_2\{\rho_{m,n}\} = \{\rho_{m,n+1}\},$$

and $E_1^\#\{\rho_{m,n}\} = \{\rho_{m-1,n}\}$ for $\{\rho_{m,n}\}_{m,n=0}^\infty \in S \times S.$

Here each sequence in S has value zero for negative integer indices.

Theorems 2.1 and 2.2 will be proved after the following Lemma 6.4 and Proposition 6.5.

LEMMA 6.4. *The recursive realations for $a_{m,n}$ are true:*

$$\{a_{m,n}\} = \alpha^m J_\mu^m (E_2 - \beta J_\mu)^{m*}\{a_{0,n}\}$$

$$+ \sum_{i=0}^{m-1} (\alpha J_\mu (E_2 - J_\mu\beta)^*)^i (J_\mu\beta)^n \{a_{m-i,0}\}$$

for $n \geq m$;

$$\{a_{m,n}\} = [J_\mu(\beta + \alpha E_1^\#)]^n\{a_{m,0}\}$$

$$= J_\mu^n \sum_{i=0}^{n} \binom{n}{i} \beta^{n-i} \alpha^i E_1^{i\#}\{a_{m,0}\}$$

$$= \{J_\mu^n \sum_{i=0}^{n} \binom{n}{i} \beta^{n-i} \alpha^i a_{m-i,0}\} \quad \text{for } m \geq n.$$

PROOF. Applying $(E_2 - J_\mu\beta)$ and $E_1^\#$, respectively, to (6.1), the relations for $a_{m,n}$ become

$$\{a_{m+1,n}\} = E_1\{a_{m,n}\}$$
$$= \alpha J_\mu (E_2 - \beta J_\mu)^*\{a_{m,n}\} + E_1(J_\mu\beta)^n\{a_{m,0}\};$$
$$\{a_{m,n+1}\} = E_2\{a_{m,n}\} \tag{6.2}$$
$$= J_\mu(\beta + \alpha E_1^\#)\{a_{m,n}\}.$$

Here, in the way similar to Section 5, $(E_2 - \beta J_\mu)^* : S \longrightarrow S$ is defined by

$$(E_2 - \beta J_\mu)^*\{b_n\} \equiv \{\sum_{i=0}^{n-1} J_\mu^{n-(i+1)} \beta^{n-(i+1)} b_i\}$$

for a sequence $\{b_n\}_{n=0}^{\infty} \in S$, and is readily seen to commute with E_1 and satisfy

$$(E_2 - \beta J_\mu)^*(E_2 - \beta J_\mu)\{b_n\} = \{b_n - J_\mu^n \beta^n b_0\};$$
$$(E_2 - \beta J_\mu)(E_2 - \beta J_\mu)^*\{b_n\} = \{b_n\}.$$

Thus, $(E_2 - \beta J_\mu)^*$ acts approximately as the inverse of $(E_2 - \beta J_\mu)$, and

$$(E_2 - \beta J_\mu)^{m*} \equiv ((E_2 - \beta J_\mu)^*)^m$$
$$\text{and } E_1^{i\#} \equiv (E_1^{\#})^i$$

for $m \in \mathbb{N} \cup \{0\}$ are defined in an obvious way.

The results follow from the recursive relations in (6.2). $\qquad \square$

Thus, as in Proposition 6.2, we have the estimate

PROPOSITION 6.5.

$$\|a_{m,n}\|$$
$$\leq M^3[(n\mu - m\lambda) + \sqrt{(n\mu - m\lambda)^2 + n\mu(\lambda - \mu)}]\gamma^n|Bx|$$
$$+ M^3 \gamma^n (1 - \lambda\omega)^{-m} \sqrt{(n\mu - m\lambda)^2 + n\mu(\lambda - \mu)}|Bx|$$
$$\text{for } n, m \geq 0,$$

where $|Bx| = \inf\{\|y\| : y \in Bx\} = \|Bx\|$.

PROOF. The proof will be divided into two cases.

Case 1: $n \geq m$. Simple calculations show

$$(E_2 - \beta J_\mu)^{2*}\{a_{0.n}\}$$
$$= \left\{\sum_{i_1=0}^{n-1} \beta^{n-(i_1+1)} \sum_{i_0-0}^{i_1-1} J_\mu^{n-i_0-1} \beta^{i_1-(i_0+1)} a_{0,n}\right\},$$

and repeating such calculations yields immediately

$$(E_2 - \beta J_\mu)^{m*}\{a_{0,n}\}$$
$$= \left\{\sum_{i_{m-1}=0}^{n-1} \beta^{n-(i_{m-1}+1)} \dots \sum_{i_1=0}^{i_2-1} \beta^{i_2-(i_1+1)}\right.$$

$$\left. \times \sum_{i_0=0}^{i_1-1} J_\mu^{n-i_0-m} \beta^{i_1-(i_0+1)} a_{0,n}\right\}. \tag{6.3}$$

On account of (6.3), it follows that

$$\|\alpha^m J_\mu^m (E_2 - \beta J_\mu)^{m*} a_{0,n}\|$$
$$\leq M(\gamma\alpha)^m \sum_{i_{m-1}=0}^{n-1} \beta^{n-(i_{m-1}+1)} \dots \sum_{i_1=0}^{i_2-1} \beta^{i_2-(i_1+1)}$$

$$\times \sum_{i_0=0}^{i_1-1} \gamma^{n-i_0-m} \beta^{i_1-(i_0+1)} \|a_{0,n}\|$$

$$\leq M^3 (\gamma\alpha)^m (E_2 - \gamma\beta)^{m*} n\gamma^n \mu|Bx|,$$

where used were the mixture condition (B3) and the inequality

$$\|a_{0,n}\| \le \sum_{i=1}^{n} \|J_\mu^i x - J_\mu^{i-1} x\|$$

$$\le M\gamma^{n-1} n \|J_\mu x - x\| \quad \text{by (B3)}$$

$$\le M^2 \gamma^n n\mu |Bx|.$$

Similarly, it is readily deduced that

$$\| \sum_{i=0}^{m-1} (\alpha J_\mu (E_2 - J_\mu \beta)^*)^i (J_\mu \beta)^n a_{m-i,0} \|$$

$$\le \sum_{i=0}^{m-1} M(\gamma\alpha(E_2 - \gamma\beta)^*)^i (\gamma\beta)^n \|a_{m-i,0}\|$$

$$\le \sum_{i=0}^{m-1} M^3 (1 - \lambda\omega)^{-m} (\gamma\alpha)^i (E_2 - \gamma\beta)^{i*} (m - i)\lambda |Bx|.$$

Hence, the result follows if the proof of Proposition 6.2 is applied to the first inequality in Lemma 6.4.

Case 2: $m \ge n$. The second inequality in Lemma 6.4 yields, on employing the mixture condition (B3),

$$\|a_{m,n}\| = M\gamma^n \sum_{i=0}^{n} \beta^{n-i} \alpha^i \|a_{m-i,0}\|$$

$$\le M^3 \gamma^n (1 - \lambda\omega)^{-m} \sum_{i=0}^{n} \beta^{n-i} \alpha^i (m - i)\lambda |Bx|.$$

Therefore, it follows again from the proof of Proposition 6.2 that

$$\|a_{m,n}\| \le M^3 \gamma^n (1 - \lambda\omega)^{-m} (m\lambda - n\mu)|Bx|.$$

Cases 1 and 2 complete the proof. $\qquad\square$

We are now in a position to do **Proof of Theorem 2.1:**

PROOF. We divide the proof into three steps.

Step 1. (The existence, continuiy, and Lipschitz continuity of $S(t)x$) Thanks to Proposition 6.5, it follows from the proof of Theorem 2.4 that

$$S(t)x \equiv \lim_{n\to\infty} (I - \frac{t}{n} B)^{-n} x$$

exist for each $x \in \overline{D(B)}$ and for bounded $t \ge 0$. The continuity and Lipschitz continuity of $S(t)x$ in $t \ge 0$, respectively, for $x \in \overline{D(B)}$ and $x \in D(B)$, also follows from that proof, where the mixture condition (B3) was used.

Step 2. (The existence of a solution) Replace the nonlinear, multi-valued operator A by the linear operator B, and replace $\ni$ and $\in$ by $=$, in the equations

(2.1), (2.2), and (2.3). It follows from Steps 1 to 5 in the proof of Theorem 2.5 that

$$S(t)x = \lim_{n\to\infty} (I - \frac{t}{n}B)^{-n}x$$
$$= \lim_{\mu\to 0}(I - \mu B)^{-[\frac{t}{\mu}]}x \quad \text{for } x \in D(B);$$

$$\lim_{\nu\to 0} u_{i-1} = \lim_{\nu\to 0}(I - \nu B)^{-[\frac{t}{\nu}]}u_0 \tag{6.4}$$
$$= S(t)u_0 \quad \text{for each } t \in [t_{i-1}, t_i);$$

$$u^n(t) - u_0 = \int_{\tau=0}^{t} B\chi^n(\tau)\,d\tau;$$

and that $u^n(t)$ converges uniformly in $t \in [0, T]$, to $u(t) \equiv S(t)u_0$. Here $u_0 \in D(B)$ and $Bu_0 \in \overline{D(B)}$.

Because B is linear, we obtain

$$Bu_i = \frac{u_i - u_{i-1}}{\nu}$$
$$= (I - \nu B)^{-(i-1)}[\frac{(I - \nu B)^{-1} - I}{\nu}]u_0$$
$$= (I - \nu B)^{-i}(Bu_0).$$

Hence, on employing the closedness of B and the mixture condition (B3), there result

$$Bu_i \text{ or } B\chi^n(t) \longrightarrow S(t)Bu_0 = Bu(t);$$
$$\|Bu_i\| = \|B\chi^n(t)\| \leq M(1 - \nu\omega)^{-[\frac{t}{\nu}]}\|Bu_0\|,$$
$$\text{uniformly bounded.}$$

Consequently, the Lebesgue dominated convergence theorem [33] implies that the last equation in (6.4) converges to, as $n \longrightarrow \infty$,

$$u(t) - u_0 = \int_{\tau=0}^{t} Bu(\tau)\,d\tau. \tag{6.5}$$

Since $Bu(t) = S(t)Bu_0$ is continuous in t for $Bu_0 \in \overline{D(B)}$ by Step 1, the fundamental theorem of calculus applied to (6.5) yields

$$\frac{du(t)}{dt} = Bu(t) = S(t)Bu_0, \quad t > 0$$
$$u(0) = u_0. \tag{6.6}$$

Thus $u(t)$ is a solution of (1.1).

Step 3. (Uniqueness of a solution, [17, pages 481-482], [14, page 83]) Let $v(t)$ be another solution of (1.1). Then it will follow that $v(t) = S(t)u_0$, proving uniqueness. For, by using Step 1 and the mixture condition (B3), the calculations, for $\frac{t}{n} < \lambda_0$ and for $x \in \overline{D(B)}$,

$$\|S(t)x\|$$
$$\leq \|S(t)x - (I - \frac{t}{n}B)^n x\| + \|(I - \frac{t}{n}B)^{-n}x\|$$
$$\leq \|S(t)x - (I - \frac{t}{n}B)^{-n}x\| + M(1 - \frac{t}{n}\omega)^{-n}\|x\|$$

for all large n

$$\longrightarrow Me^{t\omega}\|x\| \quad \text{as } n \longrightarrow \infty$$

show that $S(t)$ is a bounded operator on $\overline{D(B)}$. Hence the product rule for differentiation can be used to give, for $0 \le s \le t$,

$$\frac{d}{ds}[S(t-s)v(s)]$$

$$= (-1)S(t-s)Bv(s) + S(t-s)\frac{d}{ds}v(s)$$

$$= 0.$$

Thus $S(t-s)v(s)$ is constant in s, from which $v(t) = S(t)v(0) = S(t)u_0$ results, on setting $s = t$ and $s = 0$, respectively.   □

Finally, **Proof of Theorem 2.2:**

PROOF. Following the proof of Theorem 2.1, the results are a consequence of applying Step 1 in the proof of Theorem 2.1 to (6.6).   □

6.3. The Linear Nonhomogeneous Case. Use the quantities in Subsection 6.2:

$$J_\mu^n x = (I - \mu B)^{-n} x; \quad J_\lambda^m x = (I - \lambda B)^{-m} x;$$

$$\lambda \ge \mu > 0, \lambda\omega < 1; \quad \alpha, \beta > 0, \alpha + \beta = 1.$$

Here $x \in \overline{D(B)}$.

We begin **Proof of Theorem 2.7:**

PROOF. We divide the proof into five steps.

Step 1. As a nonlinear operator, $\tilde{B}$ satisfies the range condition (A1). For, the equation

$$u - \lambda\tilde{B}u = w,$$

where w is a given element in $\overline{D(\tilde{B})} = \overline{D(B)}$, is the same as the equation

$$u - \lambda Bu = w + \lambda f_0,$$

and f_0 satisfies the condition (F0). Here $\lambda < \lambda_0$.

Step 2. Since the linear operator B satisfies the dissipativity condition (B2) or the mixture condition (B3), it follows from Step 1 that $(I - \lambda\tilde{B})^{-1}w$ is single-valued for $w \in \overline{D(B)}$, and that

$$(I - \lambda\tilde{B})^{-1}w_2 - (I - \lambda\tilde{B})^{-1}w_1$$

$$= (I - \lambda B)^{-1}(w_2 - w_1)$$

for $w_1, w_2 \in \overline{D(B)}$. Here $0 < \lambda < \lambda_0$. Thus, a corollary of this is that, if B is dissipative, then so is $\tilde{B}$.

Step 3. Let

$$\tilde{a}_{m,n} = \tilde{J}_\mu^n x - \tilde{J}_\lambda^m x$$

for $\mu \le \lambda < \lambda_0$ and for $x \in D(B)$. Thanks to Step 2 and the resolvent identity in Lemma 4.3, we have

$$\tilde{a}_{m,n} = \tilde{J}_\mu^n x - \tilde{J}_\mu(\alpha\tilde{J}_\lambda^{m-1}x + \beta\tilde{J}_\lambda^m x)]$$

$$= J_\mu\alpha\tilde{a}_{m-1,n-1} + J_\mu\beta\tilde{a}_{m,n-1}.$$

Thus Lemma 6.4 and Proposition 6.5 and then Step 1 in the proof of Theorem 2.1 can be carried over to show the existence, continuity, and Lipschitz continuity of

$$\tilde{S}(t)x \equiv \lim_{n \to \infty} (I - \frac{t}{n}\tilde{B})^{-n}x$$
$$= \lim_{\nu \to 0} (I - \nu\tilde{B})^{-[\frac{t}{\nu}]}x.$$

Step 4. Although $\tilde{B}$ ia a nonlinear operator, we have, for $u_0 \in D(B)$ with $Bu_0 \in \overline{D(B)}$, for $0 < \nu < \lambda_0$, and for $f_0 \in \overline{D(B)}$,

$$Bu_i + f_0 = \tilde{B}u_i = \frac{u_i - u_{i-1}}{\nu}$$
$$= \frac{1}{\nu}(I - \nu\tilde{B})^{-1}u_{i-1} - \frac{1}{\nu}(I - \nu\tilde{B})^{-1}u_{i-2}$$
$$= \frac{1}{\nu}(I - \nu B)^{-1}(u_{i-1} - u_{i-2})$$
$$= \frac{1}{\nu}(I - \nu B)^{-(i-1)}(u_1 - u_0)$$
$$= (I - \nu B)^{-i}(\tilde{B}u_0) = (I - \nu B)^{-[\frac{t}{\nu}]}(Bu_0 + f_0)$$
$$\longrightarrow S(t)(Bu_0 + f_0) \quad \text{as } \nu \longrightarrow 0,$$

as is the case for B. Hence, Step 2 in the proof of Theorem 2.1 can be carried over to obtain a classical solution $\tilde{u}(t) \equiv \tilde{S}(t)u_0$ to the linear, nonhomogeneous equation (1.5).

Step 5. (Uniqueness of a solution) Let v_1 and v_2 be two solutions of the equation (1.5). Then, by substraction,

$$\frac{d}{dt}(v_1 - v_2) = \tilde{B}v_1 - \tilde{B}v_2$$
$$= B(v_1 - v_2), \quad t > 0;$$
$$(v_1 - v_2)(0) = 0,$$

so

$$v_1 \equiv v_2 \equiv \tilde{u}(t) \equiv \tilde{S}(t)u_0$$

by uniqueness of a solution for equation (1.1). ☐

We next present **Proof of Theorem 2.8:**

PROOF. In view of

$$\frac{d}{dt}\tilde{u}(t) = \tilde{B}\tilde{u}(t) = S(t)(Bu_0 + f_0),$$

a consequence of Step 4 in the above proof of Theorem 2.7, the proof is an immediate consequence of the proof of Theorem 2.2. ☐

6.4. The Nonlinear Nonhomogeneous Case. The proof of Theorems 2.10 and 2.11 will follow at once from the proof of Theorems 2.4 and 2.5. This is because the nonlinear operator $\tilde{A}$ is readily seen to satisfy the range condition (A1) and the dissipativity condition (A2).

The material of this section is based on our articles [**20, 21**].

CHAPTER 2

Existence Theorems for Evolution Equations (I)

1. Introduction

In this chapter, nonlinear evolution equations, which extend those in Chapter 1, will be investigated. In a way similar to Chapter 1, those equations will be solved again with the aid of elementary difference equations. The obtained results will be applied to solve simple, initial-boundary value problems for parabolic, partial differential equations with time-dependent coefficients. More applications to solving more general, parabolic partial differential equations with time-dependent coefficients will be given in Chapters 5 and 6.

Let $(X, \| \cdot \|)$ be a real Banach space with the norm $\| \cdot \|$, and let $T > 0$ and ω be two real constants. Consider the nonlinear evolution equation

$$\frac{du(t)}{dt} \in A(t)u(t), \quad 0 \le s < t < T,$$
$$u(s) = u_0, \tag{1.1}$$

where

$$A(t) : D(A(t)) \subset X \longrightarrow X$$

is a nonlinear, time-dependent, and multi-valued operator. Since $A(t)$ depends on t and is multi-valued, this equation extends those in Chapter 1.

To solve the evolution equation (1.1), let its approximate equation, a difference equation, be looked at first [8, Page 72]

$$\frac{u_\epsilon(t) - u_\epsilon(t - \epsilon)}{\epsilon} \in A(t)u_\epsilon(t), \quad 0 \le s < t < T,$$
$$u_\epsilon(s) = u_0. \tag{1.2}$$

Here $\epsilon > 0$ is very small. This approximate equation (1.2) is reduced to the equation

$$u_\epsilon(t) = [I - \epsilon A(t)]^{-1} u_\epsilon(t - \epsilon), \quad 0 \le s < t < T,$$
$$u_\epsilon(s) = u_0, \tag{1.3}$$

if the quantity

$$J_\epsilon(t) \equiv [I - \epsilon A(t)]^{-1}$$

can be defined. Here I is the identity operator. Thus when $\epsilon = \frac{t-s}{n}$ for $n \in \mathbb{N}$, the equation (1.3) is readily seen to have the solution

$$u_\epsilon(t) = [I - \epsilon A(s + n\epsilon)]^{-1} [I - \epsilon A(s + (n-1)\epsilon)]^{-1} \cdots$$
$$[I - \epsilon A(s + \epsilon)]^{-1} u_0$$
$$\equiv \prod_{i=1}^{n} [I - \epsilon A(s + i\epsilon)]^{-1} u_0,$$

provided that the quantity

$$\prod_{i=1}^{n} J_\epsilon(s + i\epsilon)u_0 \equiv \prod_{i=1}^{n}[I - \epsilon A(s + i\epsilon)]^{-1}u_0$$

can be defined for each $0 \le s < t < T$ and for each $n = 1, 2, \ldots$. As a result, the evolution equation (1.1) might be solved by taking the limit, as $n \longrightarrow \infty$, of the quantity

$$\prod_{i=1}^{n} J_\epsilon(s + i\epsilon)u_0.$$

But this will be true under suitable assumptions on the evolution operator $A(t)$, as the following describes it.

Equation (1.1) will be solved under the set of hypotheses, namely, the dissipativity condition (H1), the range condition (H2), and the time-regulating condition (HA) or $(HA)'$.

(H1) For each $0 \le t \le T$, $A(t)$ is dissipative in the sense that

$$\|u - v\| \le \|(u - v) - \lambda(g - h)\|$$

for all $u, v \in D(A(t))$, $g \in (A(t) - \omega)u$, $h \in (A(t) - \omega)v$, and for all $\lambda > 0$.

(H2) The range of $(I - \lambda A(t))$, denoted by E, is independent of t and contains $\overline{D(A(t))}$ for all $t \in [0, T]$ and for small $0 < \lambda < \lambda_0$, where λ_0 is some positive number satisfying $\lambda_0\omega < 1$.

(HA) There is a continuous function $f : [0, T] \longrightarrow \mathbb{R}$, of bounded variation, and there is a nonnegative function L on $[0, \infty)$ with $L(s)$ bounded for bounded s, such that, for each $0 < \lambda < \lambda_0$, we have

$$\{J_\lambda(t)x - J_\lambda(\tau)y : 0 \le t, \tau \le T, x, y \in E\} = S_1(\lambda) \cup S_2(\lambda).$$

Here $S_1(\lambda)$ denotes the set:

$$\{J_\lambda(t)x - J_\lambda(\tau)y : 0 \le t, \tau \le T, x, y \in E, \|J_\lambda(t)x - J_\lambda(\tau)y\|$$
$$\le L(\|J_\lambda(\tau)y\|)|t - \tau|\},$$

while $S_2(\lambda)$ denotes the set:

$$\{J_\lambda(t)x - J_\lambda(\tau)y : 0 \le t, \tau \le T, x, y \in E, \|J_\lambda(t)x - J_\lambda(\tau)y\|$$
$$\le (1 - \lambda\omega)^{-1}[\|x - y\|$$
$$+ \lambda|f(t) - f(\tau)|L(\|J_\lambda(\tau)y\|)(1 + \frac{\|(J_\lambda(\tau) - I)y\|}{\lambda})]\}.$$

Observe that (HA) is reduced to (H1) when $S_1(\lambda) = \emptyset$ and $t = \tau$.

$(HA)'$ There is a continuous function $f : [0, T] \longrightarrow \mathbb{R}$, of bounded variation, and there is a nonnegative function $\tilde{L}$ on $[0, \infty) \times [0, \infty)$ with $\tilde{L}(s_1, s_2)$ bounded for bounded s_1, s_2, such that, for each $0 < \lambda < \lambda_0$, we have

$$\{J_\lambda(t)x - J_\lambda(\tau)y : 0 \le t, \tau \le T, x, y \in E\} = S_1(\lambda) \cup S_2(\lambda).$$

Here $S_1(\lambda)$ denotes the set:

$$\{J_\lambda(t)x - J_\lambda(\tau)y : 0 \le t, \tau \le T, x, y \in E, \|J_\lambda(t)x - J_\lambda(\tau)y\|$$
$$\le \tilde{L}(\|J_\lambda(\tau)y\|, \|y\|)|t - \tau|\},$$

while $S_2(\lambda)$ denotes the set:

$$\{J_\lambda(t)x - J_\lambda(\tau)y : 0 \le t, \tau \le T, x, y \in E, \|J_\lambda(t)x - J_\lambda(\tau)y\|$$

$$\leq (1 - \lambda\omega)^{-1}[\|x - y\|$$
$$+ \lambda|f(t) - f(\tau)|\tilde{L}(\|J_\lambda(\tau)y\|, \|y\|)(1 + \frac{\|(J_\lambda(\tau) - I)y\|}{\lambda})]\}.$$

Again, $(HA)'$ becomes (H1), as is the case with (HA), when $S_1(\lambda) = \emptyset$ and $t = \tau$.

It is to be noted that, with (H1) and (H2) assumed, the quantity
$$J_\lambda(t)x \equiv (I - \lambda A(t))^{-1}x,$$
in $S_1(\lambda)$ or $S_2(\lambda)$ is readily seen to exist and be single-valued for $x \in E$.

The purpose of this chapter is to show that the limit
$$U(t,s)x \equiv \lim_{n\to\infty} \prod_{i=1}^{n} J_{\frac{t-s}{n}}(s + i\frac{t-s}{n})x \tag{1.4}$$

exists for $x \in \hat{D}(A(s)) = \overline{D(A(s))}$. This limit $U(t,s)x$ for $x = u_0 \in \hat{D}(A(s))$ will be only intepreted as a limit solution to equation (1.1), but it will be a strong one if $A(t)$ satisfies additionally an embedding property [20] of embeddedly quasi-demi-closedness (see Section 2). Here the quantity
$$|A(\tau)x| \equiv \lim_{\lambda\to 0} \|\frac{(J_\lambda(\tau) - I)x}{\lambda}\|$$
will exist by (H1) and (H2) and can equal ∞ [7, 8], because of Lemma 4.4 and Remark 4.5 in Chapter 1. Further, the set
$$\hat{D}(A(t)) \equiv \{x \in \overline{D(A(t))} : |A(t)x| < \infty\}$$
clearly contains the domain $D(A(t))$ of $A(t)$ and so, is called the generalized domain for $A(t)$ [7, 38].

The results of this chapter will be used in Chapters 5 and 6, where the corresponding evolution operators $A(t)$'s are second order, elliptic differential operators with time-dependent coefficients, and the corresponding evolution equations are parabolic boundary value problems with time-dependent boundary conditions.

In addition to this section, there are five more sections in this chapter. Section 2 states the main results, and Section 3 presents some simple examples. Section 4 obtains some preliminary estimates, and Section 5 proves the main results. Finally, Section 6 examines the difference equations theory in our papers [21, 22, 23], whose results, together with those in Section 4, were used in Section 5 to prove the main results in Section 2. Section 6 is a technical section with tedious calculations, so it is placed at the end of the chapter.

The material of this chapter is based on our article [25].

2. Main Results

With regard to the evolution equation (1.1), we have three theorems.

THEOREM 2.1 (Existence of a limit [25]). *Let the nonlinear operator $A(t)$ satisfy the dissipativity condition (H1), the range condition (H2), and the time-regulating condition (HA) or $(HA)'$. Then the limit*
$$U(s+t,s)u_0 \equiv \lim_{n\to\infty} \prod_{i=1}^{n} J_{\frac{t}{n}}(s + i\frac{t}{n})u_0$$

$$= \lim_{\mu \to 0} \prod_{i=1}^{[\frac{t}{\mu}]} J_\mu(s + i\mu)u_0$$

exists for $u_0 \in \hat{D}(A(s)) = \overline{D(A(s))}$ *where* $s, t \geq 0$ *and* $0 \leq (s + t) \leq T$.

This limit $U(s + t, s)u_0$ *is also continuous in* $t \geq 0$ *for* $u_0 \in \overline{D(A(s))}$, *but is Lipschitz continuous in* $t \geq 0$ *for* $u_0 \in \hat{D}(A(s))$.

Remark. Theorem 2.1 is the Crandall-Pazy theorem [**8**], if $S_1(\lambda) = \emptyset$, if $\overline{D(A(t))} \equiv \overline{D}$ is independent of t while the range of $(I - \lambda A(t))$ need not be, and if the E in (HA) is changed to $\overline{D}$.

In order to state next theorem, Theorem 2.2, concerning a limit solution and a strong solution, we need to make two preparations.

As is the case with the nonlinear, autonomous operator A in Chapter 1, the first preparation is for a limit solution. Thus, consider the discretization of (1.1) on $[0, T]$

$$u_i - \epsilon A(t_i)u_i \ni u_{i-1},$$
$$u_i \in D(A(t_i)),$$

where $n \in \mathbb{N}$ is large, and $0 < \epsilon < \lambda_0$ is such that

$$s \leq t_i = s + i\epsilon \leq T$$

for each $i = 1, 2, \ldots, n$. Here it is to be noticed that, for $u_0 \in E$, u_i exists uniquely by hypotheses (H1) and (H2).

Let $u_0 \in \hat{D}(A(s))$, and construct the Rothe functions [**12**, **32**] by defining

$$\chi^n(s) = u_0, \quad C^n(s) = A(s);$$
$$\chi^n(t) = u_i, \quad C^n(t) = A(t_i)$$
$$\text{for } t \in (t_{i-1}, t_i],$$

and

$$u^n(s) = u_0;$$
$$u^n(t) = u_{i-1} + (u_i - u_{i-1})\frac{t - t_{i-1}}{\epsilon}$$
$$\text{for } t \in (t_{i-1}, t_i] \subset [s, T].$$

It will follow (see Section 5) that, for $u_0 \in \hat{D}(A(s))$, we have

$$\lim_{n \to \infty} \sup_{t \in [0,T]} \|u^n(t) - \chi^n(t)\| = 0,$$
$$\|u^n(t) - u^n(\tau)\| \leq K_3|t - \tau|,$$
(2.1)

where $t, \tau \in (t_{i-1}, t_i]$, and that, for $u_0 \in \hat{D}A(s))$,

$$\frac{du^n(t)}{dt} \in C^n(t)\chi^n(t),$$
$$u^n(s) = u_0,$$
(2.2)

where $t \in (t_{i-1}, t_i]$. Here the last equation has values in $B([s, T]; X)$, the real Banach space of all bounded functions from $[s, T]$ to X.

The other preparation is for a strong solution, as is the case with the nonlinear, autonomous operator A in Chapter 1. Let $(Y, \|.\|_Y)$ be a real Banach space, into which the real Banach space $(X, \|.\|)$ is continuously embedded. Assume additionally that $A(t)$ satisfies the embedding condition of embeddedly quasi-demi-closedness :

(HB) If $t_n \in [0, T] \longrightarrow t$, if $x_n \in D(A(t_n)) \longrightarrow x$, and if $\|y_n\| \leq M_0$ for some $y_n \in A(t_n)x_n$ and for some positive constant M_0, then $\eta(A(t)x)$ exists and

$$|\eta(y_{n_l}) - z| \longrightarrow 0$$

for some subsequence y_{n_l} of y_n, for some $z \in \eta(A(t)x)$, and for each $\eta \in Y^* \subset X^*$, the real dual space of Y.

THEOREM 2.2 (A limit or a strong solution [**25**]). *Following Theorem 2.1, if* $u_0 \in \hat{D}(A(s))$, *then the function*

$$u(t) \equiv U(t, s)u_0 = \lim_{n \to \infty} \prod_{i=1}^{n} J_{\frac{t-s}{n}}(s + i\frac{t - s}{n})u_0$$

$$= \lim_{\mu \to 0} \prod_{i=1}^{[\frac{t-s}{\mu}]} J_\mu(s + i\mu)u_0$$

is a limit solution of the evolution equation (1.1) on $[0, T]$, in the sense that it is also the uniform limit of $u^n(t)$ on $[0, T]$, where $u^n(t)$ satisfies (2.2).

Furthermore, if $A(t)$ satisfies the embedding property (HB), then $u(t)$ is a strong solution in Y, in the sense that

$$\frac{d}{dt}u(t) \in A(t)u(t) \quad in \ Y$$

$$for \ almost \ every \ t \in (0, T);$$

$$u(s) = u_0$$

is true. The strong solution is unique if $Y \equiv X$.

It will be readily seen from the proof of Theorems 2.1 and 2.2 that

THEOREM 2.3. *The results in Theorems 2.1 and 2.2 are still true if the range condition (H2) is replaced by the weaker range condition (H2)$'$ below, provided that the initial conditions $u_0 \in \hat{D}(A(s))(\supset D(A(s)))$ and $u_0 \in \hat{D}(A(s)) = \overline{D(A(s))}(\supset D(A(s)))$ are changed to the condition $u_0 \in D(A(s))$. Here*

(H2)$'$ *The range of $(I - \lambda A(t))$, denoted by E, is independent t and contains $D(A(t))$ for all $t \in [0, T]$ and for small $0 < \lambda < \lambda_0$, where λ_0 is some positive number satisfying $\lambda_0 \omega < 1$.*

And in this case, the set $\hat{D}(A(s))$ is not well-defined, but the set

$$\tilde{D}(A(s)) \equiv \{x \in D(A(s)) : |A(s)x| < \infty\}$$

coincides with $D(A(s))$.

3. Examples

Three examples will be considered, which are time-dependent analogues of Examples 3.1, 3.2, and 3.3 in Chapter 1. The first one is about a linear, time-dependent, nonhomogeneous parabolic boundary value problem of space dimension one, and the second one is about its analogue of higher space dimensions. The last example concerns a nonlinear, time-dependent, nonhomogeneous parabolic boundary value problem of space dimension one. More complex examples will be a subject of other chapters.

EXAMPLE 3.1. Solve for $u = u(x,t)$:

$$u_t(x,t) = u_{xx}(x,t) + f_0(x,t),$$
$$(x,t) \in (0,1) \times (0,T);$$
$$u_x(0,t) = \beta_0 u(0,t), \quad u_x(1,t) = -\beta_1 u(1,t);$$
$$u(x,0) = u_0(x);$$

(3.1)

where T, β_0, and β_1 are three positive constants, and $u_t(x,t) \equiv \frac{\partial}{\partial t}u$, $u_x(x,t) \equiv \frac{\partial}{\partial x}u$, and $u_{xx}(x,t) \equiv \frac{\partial^2}{\partial x^2}u$, respectively. Here $f_0(x,t)$ is jointly continuous in x,t, and satisfies, for $x \in [0,1], t, \tau \in [0,T]$, and $\zeta(t)$, a function in t of bounded variation,

$$|f_0(x,t) - f_0(x,\tau)| \le |\zeta(t) - \zeta(\tau)|.$$

Solution. Define the time-dependent operator

$$F(t) : D(F(t)) \subset C[0,1] \longrightarrow C[0,1]$$

by $F(t)v = v'' + f_0(x,t)$ for

$$v \in D(F(t))$$
$$\equiv \{w \in C^2[0,1] : w'(j) = (-1)^j \beta_j w(j), \quad j = 0,1\}.$$

It will be shown [**25**] that $F(t)$ satisfies the four conditions, namely, the dissipativity condition (H1), the range condition (H2), the time-regulating condition (HA), and the embedding condition (HB). As a result, the quantity

$$u(t) = \lim_{n\to\infty} \prod_{i=1}^{n}[I - \frac{t}{n}F(i\frac{t}{n})]^{-1}u_0$$

$$= \lim_{\nu\to 0} \prod_{i=1}^{[\frac{t}{\nu}]}[I - \nu F(i\nu)]^{-1}u_0$$

exists for $u_0 \in \overline{\hat{D}(F(0))}$, on using Theorem 2.1. If $u_0 \in \hat{D}(F(0))$, then this $u(t)$ is not only a limit solution to the equation (3.1), but even a strong one by Theorem 2.2. In the latter case, $u(t)$ also satisfies the middle equation in (3.1). Under additional assumptions on u_0 and $f_0(x,t)$, we will make further estimates, so that $u(t)$ for $u_0 \in D(F(0))$ with $F(0)u_0 = (u_0'' + f_0(x,0)) \in D(F(0))$ is, in fact, a unique classical solution.

We now begin the proof, which is composed of eight steps.

Step 1 It is readily verified as in solving Example 3.1, Chapter 1 that $F(t)$ satisfies the dissipativity condition (H1).

Step 2. From the theory of ordinary differential equations [**5**], [**24**, Corollary 2.13, Chapter 4], the range of $(I - \lambda F(t)), \lambda > 0$, equals $C[0,1]$, so $F(t)$ satisfies the range condition (H2).

Step 3. ($F(t)$ satisfies the time-regulating condition (HA).) Let $g_i(x) \in C[0,1], i = 1,2$, and let

$$v_1 = (I - \lambda F(t))^{-1}g_1;$$
$$v_2 = (I - \lambda F(\tau))^{-1}g_2;$$

where $\lambda > 0$ and $0 \le t, \tau \le T$. Then

$$(v_1 - v_2) - \lambda(v_1 - v_2)''$$
$$= \lambda[f_0(x,t) - f_0(x,\tau)] + (g_1 - g_2),$$

and so

$$\|v_1 - v_2\|_\infty \le \|g_1 - g_2\|_\infty + \lambda \max_{x \in [0,1]} |f_0(x,t) - f_0(x,\tau)|$$
$$\le \|g_1 - g_2\|_\infty + \lambda|\zeta(t) - \zeta(\tau)|,$$

proving the condition (HA). This is because, as in solving Example 3.1, Chapter 1, the maximum principle applies, that is, there is an $x_0 \in [0,1]$ such that

$$\|v_1 - v_2\|_\infty = |(v_1 - v_2)(x_0)|,$$

that, for $x_0 \in (0,1)$,

$$(v_1 - v_2)'(x_0) = 0;$$
$$(v_1 - v_2)(x_0)(v_1 - v_2)''(x_0) \le 0,$$

and that, for $x_0 \in \{0,1\}$,

$$(v_1 - v_2)'(0) \le 0 \quad \text{or} \ge 0, \text{ according as } (v_1 - v_2)(0) > 0 \text{ or } < 0;$$
$$(v_1 - v_2)'(1) \ge 0 \quad \text{or} \le 0, \text{ according as } (v_1 - v_2)(1) > 0 \text{ or } < 0.$$

Here the boundary conditions in $D(F(t))$ make $x_0 \in \{0,1\}$ impossible.

Step 4. ($F(t)$ satisfies the embedding condition (HB).) [20] Let $t_n \in [0,T]$ converge to t, $v_n \in D(F(t_n))$ converge to v in $C[0,1]$, and $\|F(t_n)v_n\|_\infty$ be uniformly bounded. It will be shown that, for each η in the self-dual space $L^2(0,1) = (L^2(0,1))^*$, $\eta(F(t)v)$ exists and

$$|\eta(F(t_n)v_n) - \eta(F(t)v)| \longrightarrow 0.$$

Here $(C[0,1]; \|\cdot\|_\infty)$ is continuously embedded into $L^2(0,1); \|\cdot\|)$.

Since $\|v_n\|_\infty$ and $\|F(t_n)v_n\|_\infty$ are uniformly bounded, so is $\|v_n\|_{C^2[0,1]}$ by the interpolation inequality [1], [13, page 135]. Hence, by Ascoli-Arzela theorem [33], a subsequence of v_n and then itself converge in $C^1[0,1]$ to v. Also, v_n is uniformly bounded in the Hilbert space $W^{2,2}(0,1)$, whence, by the Alaoglu theorem [36], a subsequence of v_n and then itself converge weakly to v [36]. It follows that, for each $\eta \in L^2(0,1)$,

$$|\eta(F(t_n)v_n) - \eta(F(t)v)| \longrightarrow 0,$$

because

$$\left| \int_0^1 [(v_n'' - v'') + (f_0(x,t_n) - f_0(x,t))]\eta \, dx \right|$$
$$\le \left| \int_0^1 (v_n'' - v'')\eta \, dx \right| + \left| \int_0^1 [f_0(x,t_n) - f_0(x,t)]\eta \, dx \right|$$
$$\longrightarrow 0.$$

Therefore $F(t)$ satisfies the embedding condition (HB).

Step 5. ($u(t)$ for $u_0 \in \hat{D}(F(0))$ satisfies the middle equation in (3.1).) Consider the discretized equation

$$u_i - \nu F(t_i)u_i = u_{i-1},$$
$$u_i \in D(F(t_i)), \tag{3.2}$$

where $u_0 \in \hat{D}(F(0)), i = 1, 2, \ldots, n, n \in \mathbb{N}$ is large, and $\nu > 0$ is such that

$$\nu < \lambda_0 \text{ and } 0 \leq t_i = i\nu \leq T.$$

Here

$$u_i = \prod_{k=1}^{i} [I - \nu F(t_k)]^{-1} u_0$$

exists uniquely by the range condition (H2) and the dissipativity condition (H1). For convenience, we also define

$$u_{-1} = u_0 - \nu F(0)u_0.$$

Now, for each $t \in [0, T)$, we have $t \in [t_i, t_{i+1})$ for some i, so $i = [\frac{t}{\nu}]$. It follows from Theorem 2.1 that, for each above t with the corresponding i,

$$\lim_{\nu \to 0} u_i = \lim_{\nu \to 0} \prod_{k=1}^{[\frac{t}{\nu}]} [I - \nu F(t_k)]^{-1} u_0$$

$$= \lim_{n \to \infty} \prod_{k=1}^{n} [I - \frac{t}{n} F(k\frac{t}{n})]^{-1} u_0$$

$$\equiv u(t)$$

exists.

On the other hand, by utilizing Proposition 4.2 in Section 4, we have

$$\|u_i\|_\infty;$$
$$\|u_i'' + f_0(x, t_i)\|_\infty = \|F(t_i)u_i\|_\infty$$
$$= \|\frac{u_i - u_{i-1}}{\nu}\|_\infty;$$

are uniformly bounded. Those, in turn, result in a bound for $\|u_i\|_{C^2[0,1]}$ by the interpolation inequality [1], [13, page 135]. Therefore it follows from Ascoli-Arzela theorem [33] that a subsequence of u_i and then itself converge in $C^1[0, 1]$ to a limit, as $\nu \longrightarrow 0$. This limit equals $u(t)$ as shown above. Consequently, $u(t)$ satisfies the middle equation in (3.1), as u_i does so.

Step 6. (Further estimates of u_i under additional assumptions on $f_0(x, t)$, where $u_0 \in D(F(0))$ with $F(0)u_0 = (u_0'' + f_0(x, 0)) \in D(F(0))$) We assume additionally that $D_t f_0(x, t)$ and $\triangle f_0(x, t)$ exist and are continuous in x, t, and that $D_t f_0(x, t)$ satisfies, for $x \in [0, 1], t, \tau \in [0, T]$,

$$|D_t f_0(x, t) - D_t f_0(x, \tau)| \leq |\zeta(t) - \zeta(\tau)|.$$

Here $D_t f_0(x, t)$ is the partial derivative of $f_0(x, t)$ with respect to t, and $\triangle f_0(x, t)$ is the second partial derivative of $f_0(x, t)$ with respect to x.

Because of $F(0)u_0 \in D(F(0))$, the u_i in Step 5 satisfies

$$u_i - \nu[u_i'' + f_0(x, t_i)] = u_{i-1}, \quad i = 0, 1, \ldots;$$
$$u_i'(0) = \beta_0 u_i(0), \quad u_i'(1) = -\beta_1 u_i(1), \quad i = -1, 0, 1, \ldots.$$

From this, it follows, on letting $v_i = \frac{u_i - u_{i-1}}{\nu}$ for $i = 0, 1, \ldots$, that

$$v_i - \nu[v_i'' + g(x, \nu, t_i)] = v_{i-1}, \quad i = 1, 2, \ldots;$$
$$v_i'(0) = \beta_0 v_i(0), \quad v_i'(1) = -\beta_1 v_i(1), \quad i = 0, 1, \ldots;$$

where, with $t_{i-1} = t_i - \nu$,

$$g(x, \nu, t_i) = g(x, \nu, t_i, t_{i-1})$$
$$= \frac{f_0(x, t_i) - f_0(x, t_{i-1})}{\nu}.$$

Here, for convenience, we also define

$$v_{-1} = v_0 - \nu[v_0'' + g(x, \nu, t_0)];$$
$$t_{-1} = 0;$$

for which $g(x, \nu, t_0) = g(x, \nu, 0) = 0$.

Thus, either from Corollary 4.3 or from the proof of Proposition 4.1 and from both the results in Proposition 4.2 and the proof of Proposition 4.2 in Section 4, we have

$$\|v_i'' + g(x, \nu, t_i)\|_\infty = \|\frac{v_i - v_{i-1}}{\nu}\|_\infty, \quad i = 0, 1, \ldots;$$

is uniformly bounded, whence so are

$$\|v_i\|_{C^2[0,1]} = \|\frac{u_i - u_{i-1}}{\nu}\|_{C^2[0,1]}$$
$$= \|u_i'' + f_0(x, t_i)\|_{C^2[0,1]}, \quad i = 0, 1, \ldots;$$
$$\|u_i\|_{C^4[0,1]}, \quad i = 0, 1, \ldots,$$

as in Step 5. This is because those v_i's above, $i = -1, 0, 1, \ldots$, satisfy the conditions (C1), (C2), and (C3) in Corollary 4.3, that is, the conditions ((4.3) or (4.4)), ((4.5) or (4.6)), and ((4.7) or (4.8)) in Section 4. A proof of it follows from applying the maximum principle argument in Step 3 and the fact that the quantity $\frac{\|u_i - u_{i-1}\|_\infty}{\nu}$ in Step 5 is bounded.

Step 7. (Existence of a solution) Now that, from Step 6, $\|u_i\|_{C^4[0,1]}, i = 2, 3, \ldots$, is uniformly bounded, it follows from the Ascoli-Arzela theorem [**33**], as in Step 5, that a subsequence of u_i and then itself, through the discretized equation (3.2), converge in $C^3[0,1]$ to the limit $u(t)$, as $\nu \longrightarrow 0$. Therefore $u(t)$ is a classical solution.

Step 8. (Uniqueness of a solution) This proceeds as in Step 5 in the proof of Example 3.2, Chapter 1.

The proof is complete. $\square$

EXAMPLE 3.2. Solve for u = u(x, t):

$$u_t(x, t) = \triangle u(x, t) + f_0(x, t),$$
$$(x, t) \in \Omega \times (0, T);$$
$$\frac{\partial}{\partial \hat{n}} u(x, t) + \beta_2(x, t) u(x, t) = 0, \quad x \in \partial\Omega; \tag{3.3}$$
$$u(x, 0) = u_0(x).$$

Here $T > 0$, and Ω is a bounded, smooth domain in $\mathbb{R}^N$; $N \geq 2$ is a positive integer, and $x = (x_1, x_2, \ldots, x_N)$; $\triangle u = \sum_{i=1}^N \frac{\partial^2}{\partial x_i^2} u$, and $u_t = \frac{\partial}{\partial t} u$; $\partial\Omega$ is the boundary of Ω, and $\frac{\partial}{\partial \hat{n}} u$ is the unit, outer, normal derivative of u; $f_0(x, t)$ is in $C^\mu(\overline{\Omega}), 0 < \mu < 1$, for all t, jointly continuous in x, t, and satisfies, for $x \in \overline{\Omega}, t, \tau \in [0, T]$, and $\zeta(t)$, a function in t of variation,

$$|f_0(x, t) - f_0(x, \tau)| \leq |\zeta(t) - \zeta(\tau)|;$$

and finally, $\beta_2(x,t)$ is in $C^{1+\mu}(\overline{\Omega})$ for all t, jointly continuous in x,t, and satisfies, for $x \in \overline{\Omega}, t, \tau \in [0,T]$, and δ, M_0, positive constants,

$$\beta_2(x,t) \geq \delta;$$
$$|\beta_2(x,t) - \beta_2(x,\tau)| \leq M_0|t - \tau|.$$

Solution. Define the time-dependent operator

$$G(t) : D(G(t)) \subset C(\overline{\Omega}) \longrightarrow C(\overline{\Omega})$$

by $G(t)v = \triangle v + f_0(x,t)$ for

$$v \in D(G(t))$$

$$\equiv \{w \in C^{2+\mu}(\overline{\Omega}) : \frac{\partial w}{\partial \hat{n}} + \beta_2(x,t)w = 0 \quad \text{for } x \in \partial\Omega\}.$$

Here $0 < \mu < 1$, is a constant.

It will be shown [**25**] that $G(t)$ satisfies the four conditions, namely, the dissipativity condition (H1), the weaker range condition $(H2)'$, the time-regulating condition (HA), and the embedding condition (HB). As a result, the quantity

$$u(t) = \lim_{n \to \infty} \prod_{i=1}^{n}[I - \frac{t}{n}G(i\frac{t}{n})]^{-1}u_0$$

$$= \lim_{\nu \to 0} \prod_{i=1}^{[\frac{t}{\nu}]}[I - \nu G(i\nu)]^{-1}u_0$$

exists for $u_0 \in \overline{\hat{D}(G(0))}$, on using Theorem 2.1. If $u_0 \in \hat{D}(G(0))$, then this $u(t)$ is not only a limit solution to the equation (3.1), but even a strong one by Theorem 2.2. In the later case, $u(t)$ also satisfies the middle equation in (3.1). Under additional assumptions on u_0 and $f_0(x,t)$, we will make further estimates, so that $u(t)$ for $u_0 \in D(G(0))$ with $G(0)u_0 = (\triangle u_0 + f_0(x,0)) \in D(G(0))$ is, in fact, a unique classical solution.

We now begin the proof, which is composed of eight steps.

Step 1 It is readily verified as in solving Example 3.2, Chapter 1 that $G(t)$ satisfies the dissipativity condition (H1).

Step 2. From the theory of linear, elliptic partial differential equations [**13**], the range of $(I - \lambda G(t)), \lambda > 0$, equals $C^{\mu}(\overline{\Omega})$, so $G(t)$ satisfies the weaker range condition $(H2)'$ because of $C^{\mu}(\overline{\Omega}) \supset D(G(t))$ for all t.

Step 3. ($G(t)$ satisfies the time-regulating condition (HA).) Let $g_i(x) \in C^{\mu}(\overline{\Omega}), i = 1, 2$, and let

$$v_1 = (I - \lambda G(t))^{-1}g_1;$$
$$v_2 = (I - \lambda G(\tau))^{-1}g_2;$$

where $\lambda > 0$ and $0 \leq t, \tau \leq T$. Then

$$(v_1 - v_2) - \lambda\triangle(v_1 - v_2)$$
$$= \lambda[f_0(x,t) - f_0(x,\tau)] + (g_1 - g_2), \quad x \in \Omega;$$
$$\frac{\partial(v_1 - v_2)}{\partial\hat{n}} + \beta_2(x,t)(v_1 - v_2) = -(\beta_2(x,t) - \beta_2(x,\tau))v_2, \quad x \in \partial\Omega;$$

so the condition (HA) follows:

$$\|v_1 - v_2\|_\infty \leq \|g_1 - g_2\|_\infty + \lambda \max_{x \in [0,1]} |f_0(x,t) - f_0(x,\tau)|$$

$$\leq \|g_1 - g_2\|_\infty + \lambda|\zeta(t) - \zeta(\tau)|; \quad \text{or}$$

$$\|v_1 - v_2\|_\infty \leq \frac{1}{\delta} M_0 \|v_2\|_\infty |t - \tau|.$$

This is because, as in proving the dissipativity condition (H1) in Step 1, the maximum principle argument applies, that is, there is an $x_0 \in \overline{\Omega}$ such that

$$\|v_1 - v_2\|_\infty = |(v_1 - v_2)(x_0)|,$$

that, for $x_0 \in \Omega$,

$$D(v_1 - v_2)(x_0) = 0;$$
$$(v_1 - v_2)(x_0) \triangle (v_1 - v_2)(x_0) \leq 0,$$

and that, for $x_0 \in \partial\Omega$,

$$\frac{\partial(v_1 - v_2)}{\partial \hat{n}}(x_0) \geq 0 \quad \text{or} \leq 0 \text{ according as } (v_1 - v_2)(x_0) > 0 \text{ or } < 0.$$

Step 4. ($G(t)$ satisfies the embedding condition (HB).) [20] Let $t_n \in [0, T]$ converge to t, $v_n \in D(G(t_n))$ converge to v in $C(\overline{\Omega})$, and $\|G(t_n)v_n\|_\infty$ be uniformly bounded. It will be shown that, for each η in the self-dual space $L^2(\Omega) = (L^2(\Omega))^*$, $\eta(G(t)v)$ exists and

$$|\eta(G(t_n)v_n) - \eta(G(t)v)| \longrightarrow 0.$$

Here $(C(\overline{\Omega}); \|\cdot\|_\infty)$ is continuously embedded into $L^2(\Omega); \|\cdot\|)$.

Since $\|v_n\|_\infty$ and $\|G(t_n)v_n\|_\infty$ are uniformly bounded, so is $\|v_n\|_{C^{1+\lambda}(\overline{\Omega})}$ for any $0 < \lambda < 1$, using the proof of (4.1) in Chapter 5. (Alternatively, uniformly bounded are $\|v_n\|_{W^{2,p}(\Omega)}$ for any $p \geq 2$ and then $\|v_n\|_{C^{1+\lambda}(\overline{\Omega})}$ for any $0 < \lambda < 1$, on using the L^p elliptic estimates [37] and the Sobolev embedding theorem [1, 13].) Hence, by the Ascoli-Arzela theorem [33], a subsequence of v_n and then itself converge in $C^1(\overline{\Omega})$ to v. Also, v_n is uniformly bounded in the Hilbert space $W^{2,2}(\Omega)$, whence, by the Alaoglu theorem [36], a subsequence of v_n and then itself converge weakly to v [36]. It follows that, for each $\eta \in L^2(\Omega)$,

$$|\eta(G(t_n)v_n) - \eta(G(t)v)| \longrightarrow 0,$$

because

$$|\int_\Omega [(\triangle v_n - \triangle v)\eta + (f_0(x, t_n) - f_0(x, t))]\eta \, dx|$$

$$\leq |\int_\Omega (\triangle v_n - \triangle v)\eta \, dx| + |\int_0^1 [f_0(x, t_n) - f_0(x, t)]\eta \, dx|$$

$$\longrightarrow 0.$$

Therefore $G(t)$ satisfies the embedding condition (HB).

Step 5. ($u(t)$ for $u_0 \in \hat{D}(G(0))$ satisfies the middle equation in (3.3).) Consider the discretized equation

$$u_i - \nu G(t_i)u_i = u_{i-1},$$
$$u_i \in D(G(t_i)), \tag{3.4}$$

where $u_0 \in \hat{D}(G(0)), i = 1, 2, \ldots, n$, $n \in \mathbb{N}$ is large, and $\nu > 0$ is such that

$$\nu < \lambda_0 \text{ and } 0 \leq t_i = i\nu \leq T.$$

Here

$$u_i = \prod_{k=1}^{i} [I - \nu G(t_k)]^{-1} u_0$$

exists uniquely by the range condition (H2) and the dissipativity condition (H1). For convenience, we also define

$$u_{-1} = u_0 - \nu G(0)u_0.$$

Now, for each $t \in [0, T)$, we have $t \in [t_i, t_{i+1})$ for some i, so $i = [\frac{t}{\nu}]$. It follows from Theorem 2.1 that, for each above t with the corresponding i,

$$\lim_{\nu \to 0} u_i = \lim_{\nu \to 0} \prod_{k=1}^{[\frac{t}{\nu}]} [I - \nu G(t_k)]^{-1} u_0$$

$$= \lim_{n \to \infty} \prod_{k=1}^{n} [I - \frac{t}{n} G(k\frac{t}{n})]^{-1} u_0$$

$$\equiv u(t)$$

exists.

On the other hand, by utilizing Proposition 4.2 in Section 4, we have

$$\|u_i\|_\infty;$$
$$\|\triangle u_i + f_0(x, t_i)\|_\infty = \|G(t_i)u_i\|_\infty$$
$$= \|\frac{u_i - u_{i-1}}{\nu}\|_\infty;$$

are uniformly bounded, whence so is $\|u_i\|_{C^{1+\lambda}(\overline{\Omega})}$ for any $0 < \lambda < 1$, using the proof of (4.1) in Chapter 5. (Alternatively, those, in turn, result in a bound for $\|u_i\|_{W^{2,p}(\Omega)}$ for any $p \geq 2$, by the L^p elliptic estimates [**37**]. Hence, a bound exists for $\|u_i\|_{C^{1+\eta}(\overline{\Omega})}$ for any $0 < \eta < 1$, as a result of the Sobolev embedding theorem [**1, 13**].) Therefore it follows from Ascoli-Arzela theorem [**33**] that a subsequence of u_i and then itself converge in $C^{1+\mu}(\overline{\Omega})$ to a limit, as $\nu \longrightarrow 0$. This limit equals $u(t)$ as shown above. Consequently, $u(t)$ satisfies the middle equation in (3.3), as u_i does so.

Step 6. (Further estimates of u_i under additional assumptions on $f_0(x, t)$, where $u_0 \in D(G(0))$ with $G(0)u_0 = (\triangle u_0 + f_0(x, 0)) \in D(G(0))$) There are three additional assumptions. One is that the $(2+\mu)$-th derivative of $f_0(x, t)$ with respect to x exists and is jointly continuous in x, t. The second is that the first partial derivative $D_t f_0(x, t)$ of $f_0(x, t)$ with respect to t exists and is jointly continuous in x, t, and satisfies, for $x \in \overline{\Omega}, t, \tau \in [0, T]$,

$$|D_t f_0(x, t) - D_t f_0(x, \tau)| \leq |\zeta(t) - \zeta(\tau)|.$$

The third is that $\beta_2(x, t)$ is twice continuously differentiable in t, or weakly satisfies, for $x \in \partial\Omega$, $\tau > 0$, and $t, t + \tau, t + 2\tau \in [0, T]$,

$$\left| \frac{\beta_2(x, t + 2\tau) - 2\beta_2(x, t + \tau) + \beta_2(x, t)}{\tau^2} \right| \leq M_0,$$

the second difference quotient of $\beta_2(x, t)$ in t being bounded.

Because of $G(0)u_0 \in D(G(0))$, the u_i in Step 5 satisfies

$$u_i - \nu[\triangle u_i + f_0(x, t_i)] = u_{i-1}, \quad x \in \Omega,$$
$$i = 0, 1, \ldots;$$

$$\frac{\partial}{\partial \hat{n}} u_i(x) + \beta_2(x, t_i) u_i(x) = 0, \quad x \in \partial\Omega,$$
$$i = -1, 0, 1, \ldots.$$

From this, it follows, on letting $w_i = \frac{u_i - u_{i-1}}{\nu}$ for $i = 0, 1, \ldots$, that

$$w_i - \nu[\triangle w_i + g(x, \nu, t_i)] = w_{i-1}, \quad x \in \Omega,$$
$$i = 1, 2, \ldots;$$

$$\frac{\partial w_i}{\partial \hat{n}} + \beta_2(x, t_i) w_i$$
$$= -\frac{\beta_2(x, t_i) - \beta_2(x, t_{i-1})}{\nu} u_{i-1}, \quad x \in \partial\Omega,$$
$$i = 0, 1, \ldots;$$

where, with $t_{i-1} = t_i - \nu$,

$$g(x, \nu, t_i) = g(x, \nu, t_i, t_{i-1})$$
$$= \frac{f_0(x, t_i) - f_0(x, t_{i-1})}{\nu}, \quad i = 0, 1, \ldots.$$

Here, for convenience, we also define

$$w_{-1} = w_0 - \nu[\triangle w_0 + g(x, \nu, t_0))];$$
$$t_{-1} = 0;$$

for which $g(x, \nu, t_0) = g(x, \nu, 0) = 0$.

Hence, either from Corollary 4.3 or from the proof of Proposition 4.1 and from both the results in Proposition 4.2 and the proof of Proposition 4.2 in Section 4, we have

$$\|\triangle w_i + g(x, \nu, t_i)\|_\infty = \|\frac{w_i - w_{i-1}}{\nu}\|_\infty, \quad i = 0, 1, \ldots;$$
$$\|w_i\|_{C^{1+\eta}(\overline{\Omega})}, \quad 0 < \eta < 1, \quad i = 0, 1, \ldots;$$

are uniformly bounded, as in Step 5, where $w_i = \frac{u_i - u_{i-1}}{\nu} = \triangle u_i + f_0(x, t_i)$; hence, so is

$$\|u_i\|_{C^{3+\eta}(\overline{\Omega})}, \quad i = 0, 1, \ldots$$

by the Schauder global regularity theorem [13, page 111]. This is because those w_i's, $i = -1, 0, 1, \ldots$, satisfy the conditions (C1), (C2), and (C3) in Corollary 4.3, that is, the conditions ((4.3) or (4.4)), ((4.5) or (4.6)), and ((4.7) or (4.8)) in Section 4, for which employed were both the maximum principle argument in Step 3 and the boundedness of $\frac{\|u_i - u_{i-1}\|_\infty}{\nu}$ in Step 5.

Step 7. (Existence of a solution) Now that, from Step 6, $\|u_i\|_{C^{3+\eta}(\overline{\Omega})}, i = 2, 3, \ldots$, is uniformly bounded, it follows from the Ascoli-Arzela theorem [33], as in Step 5, that a subsequence of u_i and then itself, through the discretized equation (3.4), converge in $C^{3+\mu}(\overline{\Omega})$ to the limit $u(t)$, as $\nu \longrightarrow 0$. Therefore $u(t)$ is a classical solution.

Step 8. (Uniqueness of a solution) This proceeds as in Step 5 in the proof of Example 3.2, Chapter 1.

The proof is complete. $\square$

EXAMPLE 3.3. Solve for $u = u(x,t)$:

$$
\begin{aligned}
u_t(x,t) &= u_{xx}(x,t) + f_1(x,t), \\
&\quad (x,t) \in (0,1) \times (0,T); \\
u_x(0,t) &\in (-1)^j \beta_j(u(j,t)), \quad j = 0,1; \\
u(x,0) &= u_0(x);
\end{aligned}
\tag{3.5}
$$

where $T > 0$, and β_0 and β_1 are maximal monotone graphs in $\mathbb{R} \times \mathbb{R}$; $f_1(x,t)$ is jointly continuous in x,t, and satisfies, for $x \in [0,1], t, \tau \in [0,T]$, and $\zeta(t)$, a function in t of bounded variation,

$$
|f_1(x,t) - f_1(x,\tau)| \le |\zeta(t) - \zeta(\tau)|.
$$

This is the nonlinear analogue of the problem in Example 3.1. Here a monotone graph β is a subset of $\mathbb{R} \times \mathbb{R}$ that satisfies

$$
(y_2 - y_1)(x_2 - x_1) \ge 0 \quad \text{for } y_i \in \beta(x_i), \quad i = 1,2.
$$

This β is a maximal monotone graph , if it is not properly contained in any other monotone graph. In this case, for any $\lambda > 0$,

$$
(I + \lambda\beta)^{-1} : \mathbb{R} \longrightarrow \mathbb{R}
$$

is single-valued and non-expansive [3].

Solution. Define the time-dependent operator

$$
H(t) : D(H(t)) \subset C[0,1] \longrightarrow C[0,1]
$$

by $H(t)v = v'' + f_1(x,t)$ for

$$
\begin{aligned}
v &\in D(H(t)) \\
&\equiv \{w \in C^2[0,1] : w'(j) \in (-1)^j \beta_j(w(j)), \quad j = 0,1\}.
\end{aligned}
$$

It will be shown [25] that $H(t)$ satisfies the four conditions, namely, the dissipativity condition (H1), the range condition (H2), the time-regulating condition (HA), and the embedding condition (HB). As a result, the quantity

$$
u(t) = \lim_{n \to \infty} \prod_{i=1}^{n} [I - \frac{t}{n} H(i\frac{t}{n})]^{-1} u_0
$$

$$
= \lim_{\nu \to 0} \prod_{i=1}^{[\frac{t}{\nu}]} [I - \nu H(i\nu)]^{-1} u_0
$$

exists for $u_0 \in \overline{\hat{D}(H(0))}$, on using Theorem 2.1. If $u_0 \in \hat{D}(H(0))$, then this $u(t)$ is not only a limit solution to the equation (3.5), but even a strong one by Theorem 2.2. In the latter case, $u(t)$ also satisfies the middle equation in (3.5). Under additional assumptions on $f_1(x,t)$ and β, we will make further estimates, so that $u(t)$ for $u_0 \in D(H(0))$ with $H(0)u_0 \in D(H(0))$ is, in fact, a unique classical solution.

We now begin the proof, which is composed of eight steps.

Step 1 It is readily verified as in solving Example 3.3, Chapter 1 that $H(t)$ satisfies the dissipativity condition (H1).

Step 2. That the range of $(I - \lambda H(t))$, $\lambda > 0$, equals $C[0,1]$ follows immediately as in solving Example 3.3, Chapter 1. Hence, $H(t)$ satisfies the range condition (H2).

Step 3. ($H(t)$ satisfies the time-regulating condition (HA).) Let $g_i(x) \in C[0,1]$, $i = 1, 2$, and let

$$v_1 = (I - \lambda H(t))^{-1} g_1;$$

$$v_2 = (I - \lambda H(\tau))^{-1} g_2;$$

where $\lambda > 0$ and $0 \le t, \tau \le T$. Then

$$(v_1 - v_2) - \lambda(v_1 - v_2)''$$
$$= \lambda[f_1(x,t) - f_1(x,\tau)] + (g_1 - g_2),$$

so

$$\|v_1 - v_2\|_\infty \le \|g_1 - g_2\|_\infty + \lambda \max_{x \in [0,1]} |f_0(x,t) - f_0(x,\tau)|$$
$$\le \|g_1 - g_2\|_\infty + \lambda|\zeta(t) - \zeta(\tau)|,$$

proving the condition (HA). This is because, as in solving Example 3.3, Chapter 1, the maximum principle applies, that is, there is an $x_0 \in [0,1]$ such that

$$\|v_1 - v_2\|_\infty = |(v_1 - v_2)(x_0)|,$$

that, for $x_0 \in (0,1)$

$$(v_1 - v_2)'(x_0) = 0;$$
$$(v_1 - v_2)(x_0)(v_1 - v_2)''(x_0) \le 0,$$

and that, for $x_0 \in \{0,1\}$,

$$(v_1 - v_2)'(0) < 0 \quad \text{or} \ge 0, \text{ according as } (v_1 - v_2)(0) > 0 \text{ or } < 0;$$
$$(v_1 - v_2)'(1) \ge 0 \quad \text{or} \le 0, \text{ according as } (v_1 - v_2)(1) > 0 \text{ or } < 0.$$

Here the boundary conditions in $D(H(t))$ make $x_0 \in \{0,1\}$ impossible.

Step 4. ($H(t)$ satisfies the embedding condition (HB).) [20] Let $t_n \in [0,T]$ converge to t, $v_n \in D(H(t_n))$ converge to v in $C[0,1]$, and $\|H(t_n)v_n\|_\infty$ be uniformly bounded. It will be shown that, for each η in the self-dual space $L^2(0,1) = (L^2(0,1))^*$, $\eta(H(t)v)$ exists and

$$|\eta(H(t_n)v_n) - \eta(H(t)v)| \longrightarrow 0.$$

Here $(C[0,1]; \|\cdot\|_\infty)$ is continuously embedded into $L^2(0,1); \|\cdot\|)$.

Since $\|v_n\|_\infty$ and $\|H(t_n)v_n\|_\infty$ are uniformly bounded, so is $\|v_n\|_{C^2[0,1]}$ by the interpolation inequality [1], [13, page 135]. Hence, by Ascoli-Arzela theorem [33], a subsequence of v_n and then itself converge in $C^1[0,1]$ to v. Also, v_n is uniformly bounded in the Hilbert space $W^{2,2}(0,1)$, whence, by the Alaoglu theorem [36], a subsequence of v_n and then itself converge weakly to v [36]. It follows that, for each $\eta \in L^2(0,1)$,

$$|\eta(H(t_n)v_n) - \eta(H(t)v)| \longrightarrow 0,$$

because

$$\left| \int_0^1 [(v_n'' - v'')\eta + (f_1(x,t_n) - f_0(x,t))]\eta \, dx \right|$$

$$\leq |\int_0^n (v_n'' - v'')\eta\,dx| + |\int_0^1 [f_1(x, t_n) - f_1(x, t)]\eta\,dx|$$
$$\longrightarrow 0.$$

Therefore $H(t)$ satisfies the embedding condition (HB).

Step 5. $(u(t)$ for $u_0 \in \hat{D}(H(0))$ satisfies the middle equation in (3.5).) Consider the discretized equation

$$u_i - \nu H(t_i)u_i = u_{i-1},$$
$$u_i \in D(H(t_i)), \tag{3.6}$$

where $u_0 \in \hat{D}(H(0)), i = 1, 2, \ldots, n$, $n \in \mathbb{N}$ is large, and $\nu > 0$ is such that

$$\nu < \lambda_0 \text{ and } 0 \leq t_i = i\nu \leq T.$$

Here

$$u_i = \prod_{k=1}^i [I - \nu H(t_k)]^{-1} u_0$$

exists uniquely by the range condition (H2) and the dissipativity condition (H1). For convenience, we also define

$$u_{-1} = u_0 - \nu H(0)u_0.$$

Now, for each $t \in [0, T)$, we have $t \in [t_i, t_{i+1})$ for some i, so $i = [\frac{t}{\nu}]$. It follows from Theorem 2.1 that, for each above t with the corresponding i,

$$\lim_{\nu \to 0} u_i = \lim_{\nu \to 0} \prod_{k=1}^{[\frac{t}{\nu}]} [I - \nu H(t_k)]^{-1} u_0$$
$$= \lim_{n \to \infty} \prod_{k=1}^n [I - \frac{t}{n} H(k\frac{t}{n})]^{-1} u_0$$
$$\equiv u(t)$$

exists.

On the other hand, by utilizing Proposition 4.2 in Section 4, we have

$$\|u_i\|_\infty;$$
$$\|u_i'' + f_1(x, t_i)\|_\infty = \|H(t_i)u_i\|_\infty$$
$$= \|\frac{u_i - u_{i-1}}{\nu}\|_\infty;$$

are uniformly bounded. Those, in turn, result in a bound for $\|u_i\|_{C^2[0,1]}$ by the interpolation inequality [1], [**13**, page 135]. Therefore it follows from Ascoli-Arzela theorem [**33**] that a subsequence of u_i and then itself converge in $C^1[0, 1]$ to a limit, as $\nu \longrightarrow 0$. This limit equals $u(t)$ as shown above. Consequently, $u(t)$ satisfies the middle equation in (3.5), as u_i does so.

Step 6. (Further estimates of u_i under additional assumptions on $f_1(x, t)$ and β, where $u_0 \in D(H(0))$ with $H(0)u_0 \in D(H(0))$) We make two additional assumptions. One is that $D_t f_1(x, t)$ and $\triangle f_1(x, t)$ exist and are continuous in x, t, and that $D_t f_1(x, t)$, satisfy, for $x \in [0, 1], t, \tau \in [0, T]$,

$$|D_t f_1(x, t) - D_t f_1(x, \tau)| \leq |\zeta(t) - \zeta(\tau)|.$$

Here $D_t f_1(x, t)$ is the first partial derivative of $f_1(x, t)$ with respect to t, and $\triangle f_1(x, t)$ is the second partial derivative of $f_1(x, t)$ with respect to x.

The other is that $\beta_j(x), j = 0, 1$, satisfy, for $x, \tau, y_1, y_2 \in \mathbb{R}$ with $\tau > 0$ and for some constant $\delta > 0$,

$$\min\{\frac{\beta_j(x) - \beta_j(y)}{x - y}\} \geq \delta;$$

$$\max\{|[\frac{\beta_j(x + \tau y_1 + \tau y_2) - \beta_j(x + \tau y_1)}{\tau y_2}$$

$$- \frac{\beta_j(x + \tau y_1) - \beta_j(x)}{\tau y_1}]/\tau|\}$$

$$\leq M_{x,y_1,y_2}.$$

Here M_{x,y_1,y_2} is a positive number depending on x, y_1, y_2, and bounded for finite x, y_1, and y_2. That is, the first difference quotient of β_j is strictly positive, and the second difference quotient like of β_j is bounded by M_{x,y_1,y_2}.

Because of $H(0)u_0 \in D(H(0))$, the u_i in Step 5 satisfies

$$u_i - \nu[u_i'' + f_1(x, t_i)] = u_{i-1}, \quad i = 0, 1, \ldots;$$

$$u_i'(0) \in \beta_0(u_i(0)), \quad u_i'(1) \in -\beta_1(u_i(1)),$$

$$i = -1, 0, 1, \ldots.$$

From this, it follows, on letting $v_i = \frac{u_i - u_{i-1}}{\nu}$ for $i = 0, 1, \ldots$, that

$$v_i - \nu[v_i'' + g(x, \nu, t_i)] = v_{i-1}, \quad i = 1, 2, \ldots;$$

$$v_i'(0) \in \frac{\beta_0(u_i(0)) - \beta_0(u_{i-1}(0))}{u_i(0) - u_{i-1}(0)} v_i(0),$$

$$i = 0, 1, \ldots;$$

$$v_i'(1) \in -\frac{\beta_1(u_i(1)) - \beta_1(u_{i-1}(1))}{u_i(1) - u_{i-1}(1)} v_i(1),$$

$$i = 0, 1, \ldots;$$

where, with $t_{i-1} = t_i - \nu$,

$$g(x, \nu, t_i) = g(x, \nu, t_i, t_{i-1})$$

$$= \frac{f_1(x, t_i) - f_1(x, t_{i-1})}{\nu}.$$

Here for convenience, we also define

$$v_{-1} = v_0 - \nu[v_0'' + g(x, \nu, t_0)];$$

$$t_{-1} = 0,$$

for which $g(x, \nu, t_0) = g(x, \nu, 0) = 0$.

Hence, either from Corollary 4.3 or from the proof of Proposition 4.1 and the results in and the proof of Proposition 4.2 in Section 4, we have

$$\|v_i'' + g(x, \nu, t_i)\|_\infty = \|\frac{v_i - v_{i-1}}{\nu}\|_\infty,$$

$$i = 0, 1, 2, \ldots;$$

is uniformly bounded, whence so are

$$\|v_i\|_{C^2[0,1]} = \|\frac{u_i - u_{i-1}}{\nu}\|_{C^2[0,1]}$$

$$= \|u_i'' + f_1(x, t_i)\|_{C^2[0,1]}, \quad i = 0, 1, 2, \ldots;$$

$$\|u_i\|_{C^4[0,1]}, \quad i = 0, 1, 2, \ldots,$$

as in Step 5. This is because those v_i's, $i = -1, 0, 1, \ldots$, satisfy the conditions (C1), (C2), and (C3) in Corollary 4.3, that is, the conditions ((4.3) or (4.4)), ((4.5) or (4.6)), and ((4.7) or (4.8)) in Section 4, for which used were both the maximum principle argument in Step 3 and the boundedness of $\frac{\|u_i - u_{i-1}\|_\infty}{\nu}$ in Step 5.

Step 7. (Existence of a solution) Now that, from Step 6, $\|u_i\|_{C^4[0,1]}, i = 2, 3, \ldots$, is uniformly bounded, it follows from the Ascoli-Arzela theorem [**33**], as in Step 5, that a subsequence of u_i and then itself, through the discretized equation (3.6), converge in $C^3[0,1]$ to the limit $u(t)$, as $\nu \longrightarrow 0$. Therefore $u(t)$ is a classical solution.

Step 8. (Uniqueness of a solution) This proceeds as in Step 5 in the proof of Example 3.2, Chapter 1.

The proof is complete. $\square$

4. Some Preliminary Estimates

Within this section and Sections 5 and 6, we can assume, without loss of generality, that $\omega \geq 0$ where ω is the ω in the hypothesis (H1). This is because the case $\omega < 0$ is the same as the case $\omega = 0$. This will be readily seen from the corresponding proofs.

To prove the main results, that is, Theorems 2.1 and 2.2 in Section 5, we need to make two preparations. One preparation is this section, and the other is Section 6. As is explained in Section 4 in Chapter 1, the two preparations will help to solve the associated recursive inequalityn Section 4. Once the inequality is solved, the proof of the main results follows. Section 6 is a technical section with tedious calculations, so it is placed at the end of the chapter.

PROPOSITION 4.1. *Let $A(t)$ satisfy the dissipativity condition (H1), the range condition (H2), and the time-regulating condition (HA) or (HA)', and let u_0 be in $D(A(s)) \subset E$ where $0 \leq s \leq T$. Let $0 < \epsilon < \lambda_0$ be so chosen that $0 < \epsilon\omega < 1$, and let $0 \leq t_i = s + i\epsilon \leq T$ where $i \in \mathbb{N}$. Then, under (HA),*

$$\|u_i - u_0\|$$
$$\leq \eta^i L(\|u_0\|)(i\epsilon) + [\eta^{i-1}b_1 + \eta^{i-2}b_2 + \cdots + \eta b_{i-1} + b_i] \tag{4.1}$$

and

$$\|\frac{u_i - u_{i-1}}{\epsilon}\|$$
$$\leq [(c_i c_{i-1} \cdots c_2)L(\|u_0\|) \quad or \ (c_i c_{i-1} \cdots c_3)L(\|u_1\|) \ or \ \cdots$$
$$or \ c_i L(\|u_{i-2}\|) \ or \ L(\|u_{i-1}\|)] + [(c_i c_{i-1} \cdots c_1)a_0 \tag{4.2}$$
$$+ (c_i c_{i-1} \cdots c_2)d_1 + (c_i c_{i-1} \cdots c_3)d_2 + \cdots$$
$$+ c_i d_{i-1} + d_i].$$

Here

$$u_i = \prod_{j=1}^{i} J_\epsilon(t_j)u_0 \quad exists \ uniquely$$

by the hypotheses (H1) and (H2);

$$\eta = (1 - \epsilon\omega)^{-1} > 1;$$
$$b_i = \eta\epsilon\|v_0\| + \eta\epsilon|f(t_i) - f(s)|L(\|u_0\|)(1 + \|v_0\|),$$

where v_0 is any element in $A(s)u_0$;

$c_i = \eta[1 + L(\|u_{i-1}\|)|f(t_i) - f(t_{i-1})|]$;

$d_i = \eta L(\|u_{i-1}\|)|f(t_i) - f(t_{i-1})|$;

the right sides of (4.2) are interpreted as

$[L(\|u_0\|)] + [c_1 a_0 + d_1]$ for $i = 1$;

$[c_2 L(\|u_0\|)$ or $L(\|u_1\|)] + [c_2 c_1 a_0 + c_2 d_1 + d_2]$ for $i = 2$;

$\ldots$, and so on;

and

$$a_0 = \|\frac{u_0 - u_{-1}}{\epsilon}\|,$$

where u_{-1} is defined by

$$u_0 - \epsilon v_0 = u_{-1},$$

with v_0 any element in $A(s)u_0$.

Under $(HA)'$, the results are similar, in which the quantities

$$\tilde{L}(\|u_j\|, \|u_{j-1}\|), \quad j = 0, 1, \ldots, (i-1),$$

replace the above $L(\|u_j\|)$.

PROOF. The proof below will be made under (HA), because it will be similar under $(HA)'$.

We will use the method of mathematical induction. Two cases will be considered, and for each case, we divide the proof into two steps.

Case 1, where (4.1) is considered.

Step 1. Claim that(4.1) is true for $i = 1$. This will follow from the arguments below.

If $(u_1 - u_0) \in S_1(\epsilon)$ (defined in Section 1), then

$$\begin{aligned}
\|u_1 - u_0\| &= \|J_\epsilon(t_1)u_0 - J_\epsilon(s)(I - \epsilon A(s))u_0\| \\
&\leq L(\|u_0\|)|t_1 - s| \\
&\leq L(\|u_0\|)\epsilon,
\end{aligned} \tag{4.3}$$

which is less than or equal to the right side of (4.1) with $i = 1$.

On the other hand, if $(u_1 - u_0) \in S_2(\epsilon)$ (defined in Section 1), then

$$\begin{aligned}
\|u_1 - u_0\| &\leq \eta\|u_0 - u_0\| + \eta\epsilon\|v_0\| \\
&\quad + \eta\epsilon|f(t_1) - f(s)|L(\|u_0\|)(1 + \|v_0\|),
\end{aligned} \tag{4.4}$$

which is less than or equal to the right side of (4.1) with $i = 1$. Here v_0 is any element in $A(s)u_0$.

Step 2. By assuming that (4.1) is true for $i = i - 1$, we shall show that it is also true for $i = i$.

If $(u_i - u_0) \in S_1(\epsilon)$, then

$$\begin{aligned}
\|u_i - u_0\| &= \|J_\epsilon(t_i)u_{i-1} - J_\epsilon(s)(I - \epsilon A(s))u_0\| \\
&\leq L(\|u_0\|)|t_i - s| \\
&= L(\|u_0\|)(i\epsilon),
\end{aligned} \tag{4.5}$$

which is less than or equal to the right side of (4.1) with $i = i$ because of $\eta^i > 1$.

On the other hand, if $(u_i - u_0) \in S_2(\epsilon)$, then

$$\|u_i - u_0\| \leq \eta\|u_{i-1} - u_0\| + b_i \tag{4.6}$$

where

$$\eta = (1 - \epsilon\omega)^{-1},$$
$$b_i = \eta\epsilon\|v_0\| + \eta\epsilon|f(t_i) - f(s)|L(\|u_0\|)(1 + \|v_0\|).$$

This recursive inequality, combined with the induction assumption, readily gives

$$\|u_i - u_0\|$$
$$\leq \eta\{\eta^{i-1}L(\|u_0\|)(i - 1)\epsilon$$
$$\quad + [\eta^{i-2}b_1 + \eta^{i-3}b_2 + \cdots + \eta b_{i-2} + b_{i-1}]\} + b_i$$
$$= \eta^i L(\|u_0\|)(i - 1)\epsilon + [\eta^{i-1}b_1 + \eta^{i-2}b_2 + \cdots + \eta b_{i-1} + b_i],$$

which is less than or equal to the right side of (4.1) with $i = i$ because of $(i - 1)\epsilon \leq i\epsilon$.

Case 2, where (4.2) **is considered.**

Step 1. Claim that (4.2) is true for $i = 1$. This follows from the Step 1 in Case 1, because there it was shown that

$$\|u_1 - u_0\| \leq L(\|u_0\|)\epsilon \text{ or } b_1,$$

which, when divided by ϵ, is less than or equal to the right side of (4.2) with $i = 1$. Here $a_0 = \|v_0\|$, in which $a_0 = (u_0 - u_{-1})/\epsilon$ and $u_{-1} \equiv u_0 - \epsilon v_0$.

Step 2. By assuming that (4.2) is true for $i = i - 1$, we will show that it is also true for $i = i$.

If $(u_i - u_{i-1}) \in S_1(\epsilon)$, then

$$\|u_i - u_{i-1}\| \leq L(\|u_{i-1}\|)|t_i - t_{i-1}| = L(\|u_{i-1}\|)\epsilon. \tag{4.7}$$

This, when divided by ϵ, has its right side less than or equal to one of the right sides of (4.2) with $i = i$.

If $(u_i - u_{i-1}) \in S_2(\epsilon)$, then

$$\|u_i - u_{i-1}\|$$
$$\leq (1 - \epsilon\omega)^{-1}[\|u_{i-1} - u_{i-2}\|$$
$$\quad + \epsilon|f(t_i) - f(t_{i-1})|L(\|u_{i-1}\|)(1 + \frac{\|u_{i-1} - u_{i-2}\|}{\epsilon})]. \tag{4.8}$$

By letting

$$a_i = \frac{\|u_i - u_{i-1}\|}{\epsilon},$$
$$c_i = (1 - \epsilon\omega)^{-1}[1 + L(\|u_{i-1}\|)|f(t_i) - f(t_{i-1})|], \quad \text{and}$$
$$d_i = L(\|u_{i-1}\|)(1 - \epsilon\omega)^{-1}|f(t_i) - f(t_{i-1})|,$$

it follows that

$$a_i \leq c_i a_{i-1} + d_i.$$

Here notice that

$$u_0 - \epsilon v_0 = u_{-1};$$

$$a_0 = \left\| \frac{u_0 - u_{-1}}{\epsilon} \right\| = \|v_0\|.$$

The above inequality, combined with the induction assumption, readily gives

$$
\begin{aligned}
a_i &\leq c_i\{[(c_{i-1}c_{i-2}\cdots c_2)L(\|u_0\|) \\
&\quad \text{or } (c_{i-1}c_{i-2}\cdots c_3)L(\|u_1\|) \text{ or } \cdots \\
&\quad \text{or } c_{i-1}L(\|u_{i-3}\|) \text{ or } L(\|u_{i-2}\|)] + [(c_{i-1}c_{i-2}\cdots c_1)a_0 \\
&\quad + (c_{i-1}c_{i-2}\cdots c_2)d_1 + (c_{i-1}c_{i-2}\cdots c_3)d_2 + \cdots \\
&\quad + c_{i-1}d_{i-2} + d_{i-1}]\} + d_i \\
&\leq [(c_i c_{i-1}\cdots c_2)L(\|u_0\|) \\
&\quad \text{or } (c_i c_{i-1}\cdots c_3)L(\|u_1\|) \text{ or } \cdots \\
&\quad \text{or } c_i L(\|u_{i-2}\|)] + [(c_i c_{i-1}\cdots c_1)a_0 \\
&\quad + (c_i c_{i-1}\cdots c_2)d_1 + (c_i c_{i-1}\cdots c_3)d_2 + \cdots \\
&\quad + c_i d_{i-1} + d_i],
\end{aligned}
$$

each of which is less than or equal to one of the right sides of (4.2) with $i = i$. The induction proof is now complete. $\qquad\square$

PROPOSITION 4.2. *Under the assumptions of Proposition 4.1, the following are true if u_0 is in $\hat{D}(A(s)) = \{y \in \overline{D(A(s))} : |A(s)y| < \infty\}$: with (HA) assumed,*

$$\|u_i - u_0\| \leq K_1(1 - \epsilon\omega)^{-i}(2i + 1)\epsilon$$
$$\leq K_1 e^{(T-s)\omega}(3)(T - s);$$

$$\left\| \frac{u_i - u_{i-1}}{\epsilon} \right\| \leq K_3;$$

where the constants K_1 and K_3 depend on the quantities:

$$K_1 = K_1(L(\|u_0\|), (T - s), \omega, |A(s)u_0|, K_B);$$
$$K_2 = K_2(K_1, (T - s), \omega, \|u_0\|);$$
$$K_3 = K_3(L(K_2), (T - s), \omega, \|u_0\|, |A(s)u_0|, K_B);$$
$$K_B \text{ is the total variation of } f \text{ on } [0, T].$$

With $(HA)'$ assumed, the results are similar, in which the quantities

$$\tilde{L}(\|u_0\|, \|u_{-1}\|) \text{ and } \tilde{L}(K_2, K_2)$$

replace the above $L(\|u_0\|)$ and $L(K_2)$, respectively.

PROOF. The proof below is made with (HA) assumed, because it is similar with $(HA)'$ assumed.

We divide the proof into two cases.

Case 1, where $u_0 \in D(A(s))$. It follows immediately from Proposition 4.1 that

$$\|u_i - u_0\| \leq N_1(1 - \epsilon\omega)^{-i}(2i + 1)\epsilon$$
$$\leq N_1 e^{(T-s)\omega}(3)(T - s);$$

$$\left\| \frac{u_i - u_{i-1}}{\epsilon} \right\| \leq N_3;$$

where the constants N_1 and N_3 depend on the quantities:

$$N_1 = N_1(L(\|u_0\|), (T-s), \omega, \|v_0\|, K_B);$$
$$N_2 = N_2(N_1, (T-s), \omega, \|u_0\|);$$
$$N_3 = N_3(L(N_2), (T-s), \omega, \|u_0\|, \|v_0\|, K_B);$$

K_B is the total variation of f on $[0, T]$.

Here the estimate in [**8**, Page 65]

$$c_i \cdots c_1 \leq e^{i\epsilon\omega} e^{e_i + \cdots + e_1}$$

was used, where $e_i = L(\|u_{i-1}\|)|f(t_i) - f(t_{i-1})|$.

Case 2, where $u_0 \in \hat{D}(A(s))$. This involves two steps.

Step 1. Let $u_0^\mu = (I - \mu A(s))^{-1} u_0$ where $\mu > 0$, and let

$$u_i = \prod_{j=1}^{i} J_\epsilon(t_j) u_0; \quad u_i^\mu = \prod_{j=1}^{i} J_\epsilon(t_j) u_0^\mu.$$

As in [**31**, Lemma 3.2, Page 9], we have, by letting $\mu \longrightarrow 0$,

$$u_0^\mu \longrightarrow u_0;$$

here notice that $D(A(s))$ is dense in $\hat{D}(A(s))$.

Also it is readily seen that

$$u_i^\mu = \prod_{k=1}^{i} (I - \epsilon A(t_k))^{-1} u_0^\mu \longrightarrow u_i = \prod_{k=1}^{i} (I - \epsilon A(t_k))^{-1} u_0$$

as $\mu \longrightarrow 0$, since $(A(t) - \omega)$ is dissipative for each $0 \leq t \leq T$.

Step 2. Since $u_0^\mu \in D(A(s))$, Case 1 gives

$$\|u_i^\mu - u_0^\mu\|$$
$$\leq N_1(L(\|u_0^\mu\|), (T-s), \omega, \|v_0^\mu\|, K_B)(1 - \epsilon\omega)^{-i}(2i+1)\epsilon;$$
$$\frac{\|u_i^\mu - u_{i-1}^\mu\|}{\epsilon} \quad\quad\quad\quad (4.9)$$
$$\leq N_3(L(N_2), (T-s), \omega, \|u_0^\mu\|, \|v_0^\mu\|, K_B),$$

where

$$N_2 = N_2(N_1, (T-s), \omega, \|u_0^\mu\|),$$

and v_0^μ is any element in $A(s)(I - \mu A(s))^{-1} u_0$. We can take

$$v_0^\mu = w_0^\mu \equiv \frac{(J_\mu(s) - I)u_0}{\mu},$$

since $w_0^\mu \in A(s)(I - \mu A(s))^{-1} u_0$.

On account of $u_0 \in \hat{D}(A(s))$, we have

$$\lim_{\mu \to 0} \left\| \frac{(J_\mu(s) - I)u_0}{\mu} \right\| = |A(s)u_0| < \infty.$$

Thus, by letting $\mu \longrightarrow 0$ in (4.9) and using Step 1, the results in the Proposition 4.2 follow.

The proof is complete.   □

It follows immediately from the proof of Propositions 4.1 and 4.2 that

COROLLARY 4.3. *Let $u_0 \in D(A(s))$ and let $u_i, i = 1, 2, \ldots$, satisfy the difference relation, where $0 < \epsilon < \lambda_0$,*

$$u_i - \epsilon A(t_i) u_i \ni u_{i-1}, \quad i = 1, 2, \ldots.$$

Then the conclusions in Propositions 4.1 and 4.2 are still true, if we do not assume that $A(t)$ satisfies the dissipativity condition (H1), the range condition (H2), and the time-regulating condition (HA) or $(HA)'$, but assume that u_i satisfies either the three conditions (C1), (C2), and (C3) (or equally ((4.3) or (4.4)), ((4.5) or (4.6)), and ((4.7) or (4.8))) or the three conditions $(C1)', (C2)'$, and $(C3)'$:

(C1)

$$\|u_1 - u_0\|$$
$$\leq \begin{cases} L(\|u_0\|)\epsilon, & or \\ \eta\epsilon\|v_0\| + \eta\epsilon|f(t_1) - f(s)|L(\|u_0\|)(1 + \|v_0\|). \end{cases}$$

(C2)

$$\|u_i - u_0\| \leq \begin{cases} L(\|u_0\|)(i\epsilon), & or \\ \eta\|u_{i-1} - u_0\| + b_i. \end{cases}$$

(C3)

$$\|u_i - u_{i-1}\|$$
$$\leq \begin{cases} L(\|u_{i-1}\|)\epsilon, & or \\ (1 - \epsilon\omega)^{-1}[\|u_{i-1} - u_{i-2}\| \\ \quad + \epsilon|f(t_i - f(t_{i-1})|L(\|u_{i-1}\|)(1 + \frac{\|u_{i-1} - u_{i-2}\|}{\epsilon})]. \end{cases}$$

Here

v_0 *is any element in $A(s)u_0$;*

$u_0 - \epsilon v_0 = u_{-1}$;

$\eta = (1 - \epsilon\omega)^{-1}$;

$b_i = \eta\epsilon\|v_0\| + \eta\epsilon|f(t_i) - f(s)|L(\|u_0\|)(1 + \|v_0\|)$;

functions f and L are as in (HA).

$(C1)'$

$$\|u_1 - u_0\|$$
$$\leq \begin{cases} \tilde{L}(\|u_0\|, \|u_{-1}\|)\epsilon, & or \\ \eta\epsilon\|v_0\| + \eta\epsilon|f(t_1) - f(s)|\tilde{L}(\|u_0\|, \|u_{-1}\|)(1 + \|v_0\|). \end{cases}$$

$(C2)'$

$$\|u_i - u_0\| \leq \begin{cases} \tilde{L}(\|u_0\|, \|u_{-1}\|)(i\epsilon), & or \\ \eta\|u_{i-1} - u_0\| + b_i. \end{cases}$$

$(C3)'$

$$\|u_i - u_{i-1}\|$$
$$\leq \begin{cases} \tilde{L}(\|u_{i-1}\|, \|u_{i-2}\|)\epsilon, & or \\ (1 - \epsilon\omega)^{-1}[\|u_{i-1} - u_{i-2}\| \\ \quad + \epsilon|f(t_i - f(t_{i-1})|\tilde{L}(\|u_{i-1}\|, \|u_{i-2}\|)(1 + \frac{\|u_{i-1} - u_{i-2}\|}{\epsilon})]. \end{cases}$$

Here

v_0 *is any element in* $A(s)u_0$;

$u_0 - \epsilon v_0 = u_{-1}$;

$\eta = (1 - \epsilon\omega)^{-1}$;

$b_i = \eta\epsilon\|v_0\| + \eta\epsilon|f(t_i) - f(s)|\tilde{L}(\|u_0\|, \|u_{-1}\|)(1 + \|v_0\|)$;

functions f *and* $\tilde{L}$ *are as in* $(HA)'$.

5. Proof of the Main Results

Proof of Theorem 2.1 will be done after those of Propositions 5.1 and 5.2 below.

Using the preliminary estimates in Section 4, together with the difference equations theory in Section 6, it will be shown in Section 6 that

PROPOSITION 5.1. *Under the assumptions of Proposition 5.2 with (HA) assumed, the inequality is true*

$$a_{m,n} \leq \begin{cases} L(K_2)|n\mu - m\lambda|, & \text{if } S_2(\mu) = \emptyset; \\ c_{m,n} + s_{m,n} + d_{m,n} + f_{m,n} + g_{m,n}, & \text{if } S_1(\mu) = \emptyset; \end{cases}$$

where $a_{m,n}, c_{m,n}, s_{m,n}, f_{m,n}, g_{m,n}$ *and* $L(K_2)$ *are defined in Proposition 5.2.*

With $(HA)'$ *assumed, the results are similar, in which the quantities*

$$\tilde{L}(\|u_0\|, \|u_{-1}\|) \text{ and } \tilde{L}(K_2, K_2)$$

replace the above $L(\|u_0\|)$ *and* $L(K_2)$, *respectively.*

In view of this and Proposition 4.1, we are led to the claim:

PROPOSITION 5.2. *Let* $x \in \hat{D}(A(s))$ *where* $0 \leq s \leq T$, *and let* $\lambda, \mu > 0$, $n, m \in \mathbb{N}$, *be such that* $0 \leq (s + m\lambda), (s + n\mu) \leq T$, *and such that* $\lambda_0 > \lambda \geq \mu > 0$ *for which* $\mu\omega, \lambda\omega < 1$. *If* $A(t)$ *satisfies the dissipativity condition (H1), the range condition (H2), and the time-regulating condition (HA), then the inequality is true:*

$$a_{m,n} \leq c_{m,n} + s_{m,n} + d_{m,n} + e_{m,n} + f_{m,n} + g_{m,n}. \tag{5.1}$$

Here

$$a_{m,n} \equiv \|\prod_{i=1}^{n} J_\mu(s + i\mu)x - \prod_{i=1}^{m} J_\lambda(s + i\lambda)x\|;$$

$$\gamma \equiv (1 - \mu\omega)^{-1} > 1; \quad \alpha \equiv \frac{\mu}{\lambda}; \quad \beta \equiv 1 - \alpha;$$

$$c_{m,n} = 2K_1\gamma^n[(n\mu - m\lambda) + \sqrt{(n\mu - m\lambda)^2 + (n\mu)(\lambda - \mu)}];$$

$$s_{m,n} = 2K_1\gamma^n(1 - \lambda\omega)^{-m}\sqrt{(n\mu - m\lambda)^2 + (n\mu)(\lambda - \mu)};$$

$$d_{m,n} = [K_4\rho(\delta)\gamma^n(m\lambda)] + \{K_4\frac{\rho(T)}{\delta^2}\gamma^n[(m\lambda)(n\mu - m\lambda)^2$$
$$+ (\lambda - \mu)\frac{m(m+1)}{2}\lambda^2]\};$$

$$e_{m,n} = L(K_2)\gamma^n\sqrt{(n\mu - m\lambda)^2 + (n\mu)(\lambda - \mu)};$$

$$f_{m,n} = K_1[\gamma^n\mu + \gamma^n(1 - \lambda\omega)^{-m}\lambda];$$

$$g_{m,n} = K_4\rho(|\lambda - \mu|)\gamma^n(m\lambda);$$

$$K_4 = \gamma L(K_2)(1 + K_3); \quad \delta > 0 \text{ is arbitrary;}$$

$$\rho(r) \equiv \sup\{|f(t) - f(\tau)| : 0 \le t, \tau \le T, |t - \tau| \le r\}$$

$$\text{is the modulus of continuity of } f \text{ on } [0, T];$$

and $K_1, K_2,$ and K_3 are defined in Proposition 4.2.

If, instead of (HA), $(HA)'$ is assumed, the results are similar, in which the quantities $\tilde{L}(\|u_0\|, \|u_{-1}\|)$ and $\tilde{L}(K_2, K_2)$ replace the above $L(\|u_0\|)$ and $L(K_2)$, respectively.

PROOF. The proof below will be made under (HA), because it will be similar under $(HA)'$.

We will use the method of mathematical induction and divide the proof into two steps. Step 2 will involve six cases.

Step 1. (5.1) is clearly true by Proposition 4.2, if $(m, n) = (0, n)$ or $(m, n) = (m, 0)$.

Step 2. By assuming that (5.1) is true for $(m, n) = (m - 1, n - 1)$ or $(m, n) = (m, n - 1)$, we will show that it is also true for $(m, n) = (m, n)$. This is done by the arguments below.

Using the nonlinear resolvent identity in [6], we have

$$a_{m,n} = \|J_u(s + n\mu) \prod_{i=1}^{n-1} J_\mu(s + i\mu)x$$

$$- J_\mu(s + m\lambda)[\alpha \prod_{i=1}^{m-1} J_\lambda(s + i\lambda)x + \beta \prod_{i=1}^{m} J_\lambda(s + i\lambda)x)]\|.$$

Here $\alpha = \frac{\mu}{\lambda}$ and $\beta = \frac{\lambda - \mu}{\lambda}$.

Under the time-regulating condition (HA), it follows that, if the element inside the norm of the right side of the above equality is in $S_1(\mu)$, then, by Proposition 4.2 with $\epsilon = \mu$,

$$a_{m,n} \le L(\| \prod_{i=1}^{n} J_\mu(s + i\mu)x\|)|m\lambda - n\mu| \tag{5.2}$$

$$\le L(K_2)|m\lambda - n\mu|,$$

which is less than or equal to the right side of (5.1) with $(m, n) = (m, n)$, where $\gamma^n > 1$.

If that element instead lies in $S_2(\mu)$, then, by Proposition 4.2 with $\epsilon = \mu$,

$$a_{m,n} \le \gamma(\alpha a_{m-1,n-1} + \beta a_{m,n-1})$$

$$+ \gamma\mu|f(s + m\lambda) - f(s + n\mu)|L(\| \prod_{i=1}^{n} J_\mu(s + i\mu)x\|)$$

$$\times [1 + \| \frac{\prod_{i=1}^{n} J_\mu(s + i\mu)x - \prod_{i=1}^{n-1} J_\mu(s + i\mu)x}{\mu}\|] \tag{5.3}$$

$$\le [\gamma\alpha a_{m-1,n-1} + \gamma\beta a_{m,n-1}] + K_4\mu\rho(|n\mu - m\lambda|),$$

where $K_4 = \gamma L(K_2)(1 + K_3)$ and $\rho(r)$ is the modulus of continuity of f on $[0, T]$. From this, it follows that proving the following relations is sufficient under the induction assumption:

$$\gamma\alpha p_{m-1,n-1} + \gamma\beta p_{m,n-1} \le p_{m,n}; \tag{5.4}$$

$$\gamma\alpha q_{m-1,n-1} + \gamma\beta q_{m,n-1} + K_4\mu\rho(|n\mu - m\lambda|) \leq q_{m,n}; \qquad (5.5)$$

where $q_{m,n} = d_{m,n}$, and $p_{m,n} = c_{m,n}$ or $s_{m,n}$ or $e_{m,n}$ or $f_{m,n}$ or $g_{m,n}$.

Now we consider six cases.

Case 1: where $p_{m,n} = c_{m,n}$. Under this case, (5.4) is true because of the calculations, where

$$b_{m,n} = \sqrt{(n\mu - m\lambda)^2 + (n\mu)(\lambda - \mu)}$$

was defined and the Schwartz inequality was used:

$$\alpha[(n-1)\mu - (m-1)\lambda] + \beta[(n-1)\mu - m\lambda] = (n\mu - m\lambda);$$

$$\alpha b_{m-1,n-1} + \beta b_{m,n-1} = \sqrt{\alpha}\sqrt{\alpha} b_{m-1,n-1} + \sqrt{\beta}\sqrt{\beta} b_{m,n-1}$$

$$\leq (\alpha+\beta)^{\frac{1}{2}}(\alpha b_{m-1,n-1}^2 + \beta b_{m,n-1}^2)^{\frac{1}{2}}$$

$$\leq \{(\alpha+\beta)(n\mu - m\lambda)^2 + 2(n\mu - m\lambda)[\alpha(\lambda - \mu) - \beta\mu]$$

$$+ [\alpha(\lambda-\mu)^2 + \beta\mu^2] + (n-1)\mu(\lambda - \mu)\}^{\frac{1}{2}}$$

$$= b_{m,n}.$$

Here

$$\alpha + \beta = 1; \quad \alpha(\lambda - \mu) - \beta\mu = 0; \quad \alpha(\lambda - \mu)^2 + \beta\mu^2 = \mu(\lambda - \mu).$$

Case 2: where $p_{m,n} = s_{m,n}$. Under this case, (5.4) is true, as is with the Case 1, by noting that

$$(1 - \lambda\omega)^{-(m-1)} \leq (1 - \lambda\omega)^{-m}.$$

Case 3: where $q_{m,n} = d_{m,n}$. Under this case, (5.5) is true because of the calculations:

$$\gamma\alpha d_{m-1,n-1} + \gamma\beta d_{m,n-1} + K_4\mu\rho(|n\mu - m\lambda|)$$

$$\leq \{\gamma\alpha[K_4\rho(\delta)\gamma^{n-1}(m-1)\lambda] + \gamma\beta[K_4\rho(\delta)\gamma^{n-1}(m\lambda)]\}$$

$$+ \gamma\alpha\{K_4\frac{\rho(T)}{\delta^2}\gamma^{n-1}[(m-1)\lambda((n-1)\mu - (m-1)\lambda)^2$$

$$+ (\lambda - \mu)\frac{(m-1)m}{2}\lambda^2]\}$$

$$+ \gamma\beta\{K_4\frac{\rho(T)}{\delta^2}\gamma^{n-1}[(m\lambda)((n-1)\mu - m\lambda)^2$$

$$+ (\lambda - \mu)\frac{m(m+1)}{2}\lambda^2]\}$$

$$+ K_4\mu\rho(|n\mu - m\lambda|)$$

$$= K_4\rho(\delta)\gamma^n[(\alpha+\beta)(m\lambda) - \alpha\lambda]$$

$$+ K_4\frac{\rho(T)}{\delta^2}\gamma^n\{\alpha[(n\mu - m\lambda)^2 + 2(n\mu - m\lambda)(\lambda - \mu) + (\lambda - \mu)^2](m\lambda - \lambda)$$

$$+ [\alpha(\lambda - \mu)\frac{m(m+1)}{2}\lambda^2 - \alpha(\lambda - \mu)m\lambda^2]$$

$$+ \beta[(n\mu - m\lambda)^2 - 2(n\mu - m\lambda)\mu + \mu^2](m\lambda)$$

$$+ [\beta(\lambda - \mu)\frac{m(m+1)}{2}\lambda^2]\} + K_4\mu\rho(|n\mu - m\lambda|)$$

$$\leq K_4\rho(\delta)\gamma^n[(m\lambda) - \mu] + K_4\mu\rho(|n\mu - m\lambda|)$$

$$+ K_4 \frac{\rho(T)}{\delta^2} \gamma^n [(m\lambda)(n\mu - m\lambda)^2$$
$$+ (\lambda - \mu) \frac{m(m+1)}{2} \lambda^2 - \mu(n\mu - m\lambda)^2]$$
$$\equiv r_{m,n},$$

where the negative terms $[2(n\mu - m\lambda)(\lambda - \mu) + (\lambda - \mu)^2](-\lambda)$ were dropped,

$$\alpha 2(n\mu - m\lambda)(\lambda - \mu) - \beta 2(n\mu - m\lambda)\mu = 0,$$

and

$$[\alpha(\lambda - \mu)^2 + \beta\mu^2](m\lambda) = (m\lambda)\mu(\lambda - \mu),$$

which cancelled

$$-\alpha(\lambda - \mu)m\lambda^2 = -(m\lambda)\mu(\lambda - \mu);$$

it follows that $r_{m,n} \le d_{m,n}$, since

$$K_4 \mu\rho(|n\mu - m\lambda|)$$

$$\le \begin{cases} K_4\mu\rho(\delta) \le K_4\mu\rho(\delta)\gamma^n, & \text{if } |n\mu - m\lambda| \le \delta; \\ K_4\mu\rho(T)\frac{(n\mu-m\lambda)^2}{\delta^2} \le K_4\mu\rho(T)\gamma^n \frac{(n\mu-m\lambda)^2}{\delta^2}, & \text{if } |n\mu - m\lambda| > \delta. \end{cases}$$

Case 4: where $p_{m,n} = e_{m,n}$. Under this case, (5.4) is true, as is with the Case 1.

Case 5: where $p_{m,n} = f_{m,n}$. Under this case, (5.4) is true because of the calculations:

$$\gamma\alpha f_{m-1,n-1} + \gamma\beta f_{m,n-1}$$
$$= \gamma\alpha K_1[\gamma^{n-1}\mu + \gamma^{n-1}(1 - \lambda\omega)^{-(m-1)}\lambda]$$
$$+ \gamma\beta K_1[\gamma^{n-1}\mu + \gamma^{n-1}(1 - \lambda\omega)^{-m}\lambda]$$
$$\le K_1[(\alpha + \beta)\gamma^n\mu + (\alpha + \beta)\gamma^n(1 - \lambda\omega)^{-m}\lambda]$$
$$= f_{m,n}.$$

Case 6: where $p_{m,n} = g_{m,n}$. Under this case, (5.4) is true because of the calculations:

$$\gamma\alpha g_{m-1,n-1} + \gamma\beta g_{m,n-1}$$
$$\le K_4\gamma^n\rho(|\lambda - \mu|)\alpha(m-1)\lambda$$
$$+ K_4\gamma^n\rho(|\lambda - \mu|)\beta(m\lambda)$$
$$\le K_4\gamma^n\rho(|\lambda - \mu|)(\alpha + \beta)(m\lambda)$$
$$= g_{m,n}.$$

Now the proof is complete. $\square$

We are ready for **Proof of Theorem 2.1:**

PROOF. For $x \in \hat{D}(A(s))$, it follows from Proposition 5.2, by setting $\mu = \frac{t}{n} \le \lambda = \frac{t}{m} < \lambda_0$ and $\delta^2 = \sqrt{\lambda - \mu}$, that, as $n, m \longrightarrow \infty$, $a_{m,n}$ converges to 0 uniformly for $0 \le (s+t) \le T$. Thus

$$\lim_{n\to\infty} \prod_{i=1}^n J_{\frac{t}{n}}\left(s + i\frac{t}{n}\right)x$$

exists for $x \in \hat{D}(A(s))$. This limit also exists for $x \in \overline{\hat{D}(A(s))} = \overline{D(A(s))}$, on following the familiar limiting arguments in Chapter 1 or Crandall-Pazy [8].

On the other hand, setting $\mu = \lambda = \frac{t}{n} < \lambda_0, m = [\frac{t}{\mu}]$ and setting $\delta^2 = \sqrt{\lambda - \mu}$, it follows that

$$\lim_{n \to \infty} \prod_{i=1}^{n} J_{\frac{t}{n}}(s + i\frac{t}{n})u_0 = \lim_{\mu \to 0} \prod_{i=1}^{[\frac{t}{\mu}]} J_\mu(s + i\mu)u_0. \tag{5.6}$$

Now, to show the Lipschitz property, (5.6) and Crandall-Pazy [8, Page 71] will be used. From Proposition 4.2, it is derived that

$$\|u_n - u_m\|$$
$$\leq \|u_n - u_{n-1}\| + \|u_{n-1} - u_{n-2}\| + \cdots + \|u_{m+1} - u_m\|$$
$$\leq K_3\mu(n - m) \quad \text{for} \quad x \in \hat{D}(A(s));$$
$$u_n = \prod_{i=1}^{n} J_\mu(s + i\mu)x; \quad u_m = \prod_{i=1}^{m} J_\mu(s + i\mu)x,$$

where $n = [\frac{t}{\mu}], m = [\frac{\tau}{\mu}], t > \tau$ and $0 < \mu < \lambda_0$. The proof is completed by making $\mu \longrightarrow 0$ and using (5.6). □

Proof of Theorem 2.2 will be done after those of Propositions 5.3 and 5.4 below. With regard to Proposition 5.3, we need the following setup.

Consider the discretization of (1.1) on $[0, T]$,

$$u_i - \epsilon A(t_i)u_i \ni u_{i-1},$$
$$u_i \in D(A(t_i)), \tag{5.7}$$

where $n \in \mathbb{N}$ is large, and $0 < \epsilon < \lambda_0$ is such that $s \leq t_i = s + i\epsilon \leq T$ for each $i = 1, 2, \ldots, n$. Here, to be noticed is that, for $u_0 \in E$, u_i exists uniquely by the hypotheses (H1) and (H2).

Let $u_0 \in \hat{D}(A(s))$, and construct the Rothe functions [12, 32] by defining

$$\chi^n(s) = u_0, \quad C^n(s) = A(s);$$
$$\chi^n(t) = u_i, \quad C^n(t) = A(t_i) \tag{5.8}$$
$$\text{for } t \in (t_{i-1}, t_i],$$

and

$$u^n(s) = u_0;$$
$$u^n(t) = u_{i-1} + (u_i - u_{i-1})\frac{t - t_{i-1}}{\epsilon} \tag{5.9}$$
$$\text{for } t \in (t_{i-1}, t_i] \subset [s, T].$$

Since

$$\|\frac{u_i - u_{i-1}}{\epsilon}\| \leq K_3$$

for $u_0 \in \hat{D}(A(s))$ by Proposition 4.2, it follows immediately that

PROPOSITION 5.3. For $u_0 \in \hat{D}(A(s))$, we have that

$$\lim_{n \to \infty} \sup_{t \in [0,T]} \|u^n(t) - \chi^n(t)\| = 0;$$
$$\|u^n(t) - u^n(\tau)\| \leq K_3|t - \tau|,$$

where $t, \tau \in (t_{i-1}, t_i]$, and that

$$\frac{du^n(t)}{dt} \in C^n(t)\chi^n(t);$$
$$u^n(s) = u_0,$$

where $t \in (t_{i-1}, t_i]$. Here the last equation has values in $B([s,T];X)$, the real Banach space of all bounded functions from $[s,T]$ to X.

PROPOSITION 5.4. *If $A(t)$ satisfies the assumptions in Theorem 2.1, then*

$$\lim_{n\to\infty} u^n(t) = \lim_{n\to\infty} \prod_{i=1}^{n} J_{\frac{t-s}{n}}(s + i\frac{t-s}{n})u_0$$

$$= \lim_{\mu\to 0} \prod_{i=1}^{[\frac{t-s}{\mu}]} J_\mu(s + i\mu)u_0$$

uniformly for finite $0 \le (s+t) \le T$ and for $u_0 \in \hat{D}(A(s))$.

PROOF. The asserted uniform convergence will be proved by using the Ascoli-Arzela Theorem [**33**].

Pointwise convergence will be proved first. For each $t \in [s,T)$, we have $t \in [t_i, t_{i+1})$ for some i, so $i = [\frac{t-s}{\epsilon}]$, the greatest integer that is less than or equal to $\frac{t-s}{\epsilon}$. That u_i converges is because, for each above t,

$$\begin{aligned}
\lim_{\epsilon\to 0} u_i &= \lim_{\epsilon\to 0} \prod_{k=1}^{i} (I - \epsilon A(t_k))^{-1} u_0 \\
&= \lim_{n\to\infty} \prod_{k-1}^{n} [I - \frac{t-s}{n} A(s + k\frac{t-s}{n})]^{-1} u_0
\end{aligned} \tag{5.10}$$

by (5.6), which has the right side convergent by Theorem 2.1. Since

$$\left\|\frac{u_i - u_{i-1}}{\epsilon}\right\| \le K_3$$

for $u_0 \in \hat{D}(A(s))$, we see from the definition of $u^n(t)$ that

$$\lim_{n\to\infty} u^n(t) = \lim_{\epsilon\to 0} u_i$$

$$= \lim_{n\to\infty} \prod_{i=1}^{n} J_{\frac{t-s}{n}}(s + i\frac{t-s}{n})u_0$$

for each t.

On the other hand, due to

$$\left\|\frac{u_i - u_{i-1}}{\epsilon}\right\| \le K_3$$

again, we see that $u^n(t)$ is equi-continuous in $C([s,T];X)$, the real Banach space of all continuous functions from $[s,T]$ to X. Thus it follows from the Ascoli-Arzela theorem [**33**] that, for $u_0 \in \hat{D}(A(s))$, some subsequence of $u^n(t)$ and then itself converge uniformly to some

$$u(t) = \lim_{n \to \infty} \prod_{i=1}^{n} J_{\frac{t-s}{n}}(s + i\frac{t-s}{n})u_0$$

$$\in C([s, T]; X).$$

This completes the proof.   □

Here is **Proof of Theorem 2.2:**

PROOF. That $u(t)$ is a limit solution follows from Propositions 5.3 and 5.4. That $u(t)$ is a strong solution under the embedding property (HB) follows as in the Step 5 for the proof of Theorem 2.5 in Chapter 1.   □

6. Difference Equations Theory

In this section, Proposition 5.1 in Section 5 will be proved, using the theory of difference equations [**28**]. A basic part of this theory was introduced in Section 5 in Chapter 1, but we will collect it here for easy reference. This includes its Lemma 5.1, Proposition 5.2, and Proposition 5.3 which will be frequently referred to. In fact, we will apply them to derive the four lemmas below, after which Proposition 5.1 will be proved.

We now review the basic part of difference equations theory. Let

$$\{b_n\} = \{b_n\}_{n \in \{0\} \cup \mathbb{N}} = \{b_n\}_{n=0}^{\infty}$$

be a sequence of real numbers. For such a sequence $\{b_n\}$, we further extend it by defining

$$b_n = 0 \quad \text{if } n = -1, -2, \ldots..$$

The set of all such sequences $\{b_n\}$'s will be denoted by S. Thus, if $\{a_n\} \in S$, then

$$0 = a_{-1} = a_{-2} = \cdots.$$

Define a right shift operator $E : S \longrightarrow S$ by

$$E\{b_n\} = \{b_{n+1}\} \quad \text{for } \{b_n\} \in S.$$

Similarly, define a left shift operator $E^{\#} : S \longrightarrow S$ by

$$E^{\#}\{b_n\} = \{b_{n-1}\} \quad \text{for } \{b_n\} \in S.$$

For $c \in \mathbb{R}$ and $c \neq 0$, define the operator $(E - c)^* : S \longrightarrow S$ by

$$(E - c)^*\{b_n\} = \{c^n \sum_{i=0}^{n-1} \frac{b_i}{c^{i+1}}\}$$

for $\{b_n\} \in S$. Here the first term on the right side of the equality, corresponding to $n = 0$, is zero.

One more definition is that define, for $\{b_n\} \in S$,

$$(E - c)^{i*}\{b_n\} = [(E - c)^*]^i\{b_n\}, \quad i = 1, 2, \ldots;$$

$$E^{i\#}\{b_n\} = (E^{\#})^i\{b_n\}, \quad i = 1, 2, \ldots;$$

$$(E - c)^0\{b_n\} = \{b_n\}.$$

It will follow that $(E - c)^*$ acts approximately as the inverse of $(E - c)$ in this sense

$$(E - c)^*(E - c)\{b_n\} = \{b_n - c^n b_0\}.$$

Next we extend the above notions to doubly indexed sequences. For a doubly indexed sequence $\{\rho_{m,n}\} = \{\rho_{m,n}\}_{m,n=0}^{\infty}$ of real numbers, let

$$E_1\{\rho_{m,n}\} = \{\rho_{m+1,n}\};$$
$$E_2\{\rho_{m,n}\} = \{\rho_{m,n+1}\}.$$

Thus, E_1 and E_2 are the right shift operators, which act on the first index and the second index, respectively.

It is easy to see that

$$E_1 E_2\{\rho_{m,n}\} = E_2 E_1\{\rho_{m,n}\}$$

is true.

Here are Lemma 5.1, Proposition 5.2, and Proposition 5.3 in Chapter 1, but we relabel them as Lemma 6.1, Proposition 6.2, and Proposition 6.3, respectively:

LEMMA 6.1. *Let $\{b_n\}$ and $\{d_n\}$ be in S. Then the following are true:*

$$(E - c)^*(E - c)\{b_n\} = \{b_n - c^n b_0\};$$
$$(E - c)(E - c)^*\{b_n\} = \{b_n\};$$
$$(E - c)^*\{b_n\} \leq (E - c)^*\{d_n\}, \quad if\ c > 0\ and\ \{b_n\} \leq \{d_n\}.$$

Here $\{b_n\} \leq \{d_n\}$ means $b_n \leq d_n$ for $n = 0, 1, 2, \ldots$.

PROPOSITION 6.2. *Let $\xi, c \in \mathbb{R}$ be such that $c \neq 1$ and $c \neq 0$. Let, be in S, the three sequences $\{n\}_{n=0}^{\infty}, \{c^n\}_{n=0}^{\infty}$, and $\{\xi\}_{n=0}^{\infty}$ of real numbers. Then the following identities are true:*

$$(E - c)^*\{n\} = \{\frac{n}{d} - \frac{1}{d^2} + \frac{c^n}{d^2}\};$$
$$(E - c)^*\{\xi\} = \{\frac{\xi}{d} - \frac{\xi c^n}{d}\};$$
$$(E - c)^{i*}\{c^n\} = \{\binom{n}{i} c^{n-i}\}.$$

Here $d = 1 - c$ and $i = 0, 1, 2, \ldots$.

PROPOSITION 6.3. *Let $\xi, c \in \mathbb{R}$ be such that $c \neq 1$ and $c\xi \neq 0$. Let, be in S, the three sequences $\{n\xi^n\}_{n=0}^{\infty}, \{\xi^n\}_{n=0}^{\infty}$, and $\{(c\xi)^n\}_{n=0}^{\infty}$ of real numbers. Then the identities are true:*

$$(E - c\xi)^*\{n\xi^n\} = \{(\frac{n\xi^n}{d} - \frac{\xi^n}{d^2} + \frac{c^n\xi^n}{d^2})\frac{1}{\xi}\};$$
$$(E - c\xi)^*\{\xi^n\} = \{(\frac{\xi^n}{d} - \frac{c^n\xi^n}{d})\frac{1}{\xi}\};$$
$$(E - c\xi)^{i*}\{(c\xi)^n\} = \{\binom{n}{i}(c\xi)^{n-i}\}.$$

Here $d = 1 - c$ and $i = 0, 1, 2, \ldots$.

Before we prove the Proposition 5.1, we need the following four lemmas.

LEMMA 6.4. *With (5.3) assumed, the two inequalities hold:*

$$\{a_{m,n}\}$$

$$\leq (\alpha\gamma(E_2 - \beta\gamma)^*)^m \{a_{0,n}\} + \sum_{i=0}^{m-1} (\gamma\alpha(E_2 - \gamma\beta)^*)^i \{(\gamma\beta)^n a_{m-i,0}\}$$

$$+ \sum_{j=1}^{m} (\gamma\alpha)^{j-1}((E_2 - \gamma\beta)^*)^j \{r_{m+1-j,n+1}\} \quad \text{for } n \geq m;$$

(6.1)

$$\{a_{m.n}\}$$

$$\leq \{(\beta\gamma + \gamma\alpha E_1^{\#})^n a_{m,0}\} + \{\sum_{j=0}^{n-1} (\beta\gamma + \gamma\alpha E_1^{\#})^j r_{m,n-j}\}$$

$$= \{\gamma^n \sum_{k=0}^{n} \binom{n}{k} \beta^{n-k} \alpha^k a_{m-k,0}\} + \{\sum_{j=0}^{n-1} \sum_{i=0}^{j} \gamma^j \beta^{j-i} \alpha^i r_{m-i,n-j}\}$$

(6.2)

$$\text{for } m \geq n.$$

Here $r_{m,n} = K_4 \mu \rho(|n\mu - m\lambda|)$.

PROOF. [22] The proof will be divided into two cases.
Case 1: $n \geq m$. From (5.3), we have

$$E_1(E_2 - \beta\gamma)\{a_{m,n}\} \leq \{\alpha\gamma a_{m,n}\} + E_1 E_2 \{r_{m,n}\},$$

(6.3)

and both sides of this, when applied to by $(E_2 - \beta\gamma)^*$, yield

$$E_1\{a_{m,n}\} \leq \alpha\gamma(E_2 - \beta\gamma)^*\{a_{m,n}\} + (\beta\gamma)^n E_1\{a_{m,0}\}$$
$$+ (E_2 - \beta\gamma)^*\{r_{m+1,n+1}\}.$$

Here Lemma 6.1 was used. This recursive relation gives the desired result.
Case 2: $m \geq n$. On applying $E_1^{\#}$ to both sides of (6.3), it follows that

$$E_2\{a_{m,n}\} \leq (\beta\gamma + \alpha\gamma E_1^{\#})\{a_{m,n}\} + E_2\{r_{m,n}\}.$$

Here Lemma 6.1 was used again. This recursive relation delivers the desired result. The proof is complete. □

LEMMA 6.5. *The equality is true: for $n \geq m$,*

$$((E_2 - \beta\gamma)^*)^m \{n\gamma^n\} = \{\frac{n\gamma^n}{\alpha^m}\frac{1}{\gamma^m} - \frac{m\gamma^n}{\alpha^{m+1}}\frac{1}{\gamma^m}$$

$$+ \left(\sum_{i=0}^{m-1} \binom{n}{i} \frac{\beta^{n-i}}{\alpha^{m+1-i}}(m-i)\frac{1}{\gamma^m}\right)\gamma^n\}.$$

Here γ, α and β are defined in Proposition 5.2.

PROOF. This was proved in the proof of Proposition 6.2 in Chapter 1 and in [21]. □

LEMMA 6.6. *The equality is true:*

$$((E - \beta\gamma)^*)^j \{\gamma^n\} = \{(\frac{1}{\alpha^j} - \frac{1}{\alpha^j}\sum_{i=0}^{j-1} \binom{n}{i} \beta^{n-i}\alpha^i)\gamma^{n-j}\}$$

$$= \{(\frac{1}{\alpha^j} \sum_{i=j}^{n} \beta^{n-i}\alpha^i)\gamma^{n-j}\}$$

for $j \leq n \in \mathbb{N}$. Here γ, α and β are defined in Proposition 5.2.

PROOF. This is a consequence of repeatedly applying Proposition 6.3 and carefully grouping terms [22]. □

LEMMA 6.7. The equality is true: for $n \geq m$,

$$(E - \beta\gamma)^{m*}\{n^2\gamma^n\}$$

$$= \gamma^{n-m}\{\frac{n^2}{\alpha^m} - \frac{(2m)n}{\alpha^{m+1}} + [\frac{m(m-1)}{\alpha^{m+2}} + \frac{m(1+\beta)}{\alpha^{m+2}}]$$

$$- \sum_{j=0}^{m-1}[\frac{(m-j)(m-j-1)}{\alpha^{m-j+2}} + \frac{(m-j)(1+\beta)}{\alpha^{m-j+2}}]\binom{n}{j}\beta^{n-j}\}.$$

PROOF. [23] The proof will be divided into three steps.

Step 1. Claim that

$$(E - \beta)^*\{n^2\} = \{\frac{n^2}{\alpha} - \frac{2n}{\alpha^2} + \frac{1}{\alpha^2} - \frac{1}{\alpha^2}\beta^n\}.$$

From the definition of $(E - \beta)^*$ and Proposition 6.2, it follows that

$$\{\beta^n(\frac{1}{\beta^2} + \frac{2}{\beta^3} + \cdots + \frac{n-1}{\beta^n})\} = (E - \beta)^*\{n\}$$

$$= \{\frac{n}{\alpha} - \frac{1}{\alpha^2} + \frac{1}{\alpha^2}\beta^n\}.$$

This, when differentiated with respect to β, proves the claim.

Step 2. Claim that

$$(E - \beta\gamma)^*\{n^2\gamma^n\} = \gamma^{n-1}(E - \beta)^*\{n^2\}.$$

But this is an immediate consequence of the definition of $(E - \beta\gamma)^*$.

Step 3. Step 2, combined with Step 1, yields at once

$$(E - \beta\gamma)^*\{n^2\gamma^n\} = \{[\frac{n^2\gamma^n}{\alpha} - \frac{2n\gamma^n}{\alpha^2} + \frac{1}{\alpha^2}\gamma^n - \frac{1}{\alpha^2}(\beta\gamma)^n]\frac{1}{\gamma}\}.$$

On applying $(E - \beta\gamma)^*$ to both sides and using Proposition 6.3, we obtain

$$(E - \beta\gamma)^{2*}\{n^2\gamma^n\}$$

$$= \{[\frac{n^2}{\alpha^2} - \frac{4n}{\alpha^3} + (\frac{2}{\alpha^4} + \frac{2}{\alpha^3}) - (\frac{2}{\alpha^4} + \frac{2}{\alpha^3})\beta^n - \frac{1}{\alpha^2}\binom{n}{1}\beta^n]\gamma^n\frac{1}{\gamma^2}\}.$$

This operation, when repeated, gives

$$(E - \beta\gamma)^{3*}\{n^2\gamma^n\}$$

$$= \{[\frac{n^2}{\alpha^3} - \frac{6n}{\alpha^4} + (\frac{6}{\alpha^5} + \frac{3}{\alpha^4}) - (\frac{6}{\alpha^5} + \frac{3}{\alpha^4})\beta^n$$

$$- (\frac{2}{\alpha^4} + \frac{2}{\alpha^3})\binom{n}{1}\beta^{n-1} - \frac{1}{\alpha^2}\binom{n}{2}\beta^{n-2}]\gamma^n\frac{1}{\gamma^3}\}.$$

Continuing in this way, we are led to, for $m \in \mathbb{N}$,

$$(E - \beta\gamma)^{m*}\{n^2\gamma^n\}$$

$$= \{[\frac{n^2}{\alpha^m} - \frac{(2m)n}{\alpha^{m+1}} + (\frac{\chi_m}{\alpha^{m+2}} + \frac{m}{\alpha^{m+1}})$$

$$- \sum_{i=1}^{m}(\frac{\chi_i}{\alpha^{i+2}} + \frac{i}{\alpha^{i+1}})\binom{n}{m-i}\beta^{n-(m-i)}]\gamma^n\frac{1}{\gamma^m}\}, \tag{6.4}$$

where χ_m satisfies $\chi_0 = 0$, $\chi_1 = 0$, $\chi_2 = 2$, and

$$\chi_m = \chi_{m-1} + 2(m - 1).$$

This difference equation, when solved for χ_m, gives $\chi_m = m(m-1)$. This, together with the substitution $j = m - i$, plugged into (6.4), completes the proof. □

Here is our **Proof of Proposition 5.1:**

PROOF. The proof below is made with (HA) assumed, because it is similar with $(HA)'$ assumed.

The proof will be divided into two steps, where Step 2 involves two cases.

Step 1. If $S_2(\mu) = \emptyset$, then (5.2) is true, whence

$$a_{m,n} \leq L(K_2)|n\mu - m\lambda|.$$

This is the desired estimate.

Step 2. If $S_1(\mu) = \emptyset$, then the inequality (5.3) is true, from which so are the inequalities (6.1) and (6.2) as a consequence of Lemma 6.4. There are two cases to consider.

Case 1: (6.1) **holds.** In this case, we have, by Proposition 4.2,

$$a_{0,n} \leq K_1\gamma^n(2n + 1)\mu;$$
$$a_{m-i,0} \leq K_1(1 - \lambda\omega)^{-m}[2(m - i) + 1]\lambda; \tag{6.5}$$

so Lemma 6.6 and the proof of Proposition 6.2 in Chapter 1 imply that the first two terms of the right side of the inequality (6.1) is less than or equal to

$$\{c_{m,n} + s_{m,n} + f_{m,n}\}.$$

Thus, what remains to do is to estimate the third term, denoted by $\{t_{m,n}\}$, of the right side of the inequality (6.1).

Observe that, using the subadditivity of ρ, we have

$$\{t_{m,n}\}$$

$$\leq \sum_{j=1}^{m}(\gamma\alpha)^{j-1}(E_2 - \gamma\beta)^{j*}K_4\mu\{\rho(|\lambda - \mu|) + \rho(|n\mu - m\lambda + j\lambda|)\}$$

$$\leq \sum_{j=1}^{m}(\gamma\alpha)^{j-1}(E_2 - \gamma\beta)^{j*}K_4\mu\{\gamma^n\rho(|\lambda - \mu|) + \gamma^n\rho(|n\mu - (m - j)\lambda|)\}$$

$$\equiv \{u_{m,n}\} + \{v_{m,n}\},$$

where $\gamma = (1 - \mu\omega)^{-1} > 1$. It follows from Lemma 6.6 that

$$\{u_{m,n}\}$$

$$\leq \{K_4\mu\gamma^n\rho(|\lambda - \mu|)\sum_{j=1}^{m}\alpha^{j-1}\frac{1}{\alpha^j}\sum_{i=1}^{n}\binom{n}{i}\beta^{n-i}\alpha^i\}$$

$$\leq \{K_4 \gamma^n \rho(|\lambda - \mu|) \mu \frac{1}{\alpha} m\} = \{K_4 \rho(|\lambda - \mu|) \gamma^n (m\lambda)\}$$
$$\leq \{g_{m,n}\}.$$

To estimate $\{v_{m,n}\}$, let, as in Crandall-Pazy [8, page 68], $\delta > 0$ be given and write $\{v_{m,n}\} = \{I^{(1)}_{m,n}\} + \{I^{(2)}_{m,n}\}$, where $\{I^{(1)}_{m,n}\}$ is the sum over indices with $|n\mu - (m - j)\lambda| < \delta$, but $\{I^{(2)}_{m,n}\}$ is the sum over indices with $|n\mu - (m - j)\lambda| \geq \delta$. As a consequence of Lemma 6.6, we have

$$\{I^{(1)}_{m,n}\} \leq \{K_4 \mu \gamma^n \rho(\delta) \sum_{j=1}^{m} \alpha^{j-1} \frac{1}{\alpha^j} \sum_{i=j}^{n} \binom{n}{i} \beta^{n-i} \alpha^i\}$$

$$\leq \{K_4 \rho(\delta) \mu \gamma^n m \frac{1}{\alpha}\} = \{K_4 \rho(\delta) \gamma^n m\lambda\}.$$

On the other hand, we have

$$\{I^{(2)}_{m,n}\} \leq K_4 \mu \rho(T) \sum_{j=1}^{m} (\gamma\alpha)^{j-1} (E_2 - \gamma\beta)^{j*} \{\gamma^n\}$$

$$\leq K_4 \mu \rho(T) \sum_{j=1}^{m} (\gamma\alpha)^{j-1} (E_2 - \gamma\beta)^{j*} \{\gamma^n \frac{[n\mu - (m - j)\lambda]^2}{\delta^2}\},$$

which will be less than or equal to

$$\{K_4 \frac{\rho(T)}{\delta^2} \gamma^n [(m\lambda)(n\mu - m\lambda)^2 + (\lambda - \mu) \frac{m(m + 1)}{2} \lambda^2]\},$$

thus deriving the desired estimate

$$\{t_{m,n}\} = \{u_{m,n}\} + \{v_{m,n}\} \leq \{g_{m,n}\} + \{d_{m,n}\}.$$

This is because of the following five calculations, where Lemmas 6.5, 6.6, and 6.7 were used.

Calculation 1.

$$[n\mu - (m - j)\lambda]^2 = n^2\mu^2 - 2(n\mu)(m - j)\lambda + (m - j)^2\lambda^2.$$

Calculation 2.

$$\sum_{j=1}^{m} (\gamma\alpha)^{j-1} (E_2 - \gamma\beta)^{j*} \{\gamma^n n^2\} \mu^2$$

$$= \gamma^{n-1} \sum_{j=1}^{m} \alpha^{j-1} \{\frac{n^2}{\alpha^j} - \frac{2jn}{\alpha^{j+1}} + [\frac{j(j - 1)}{\alpha^{j+2}} + \frac{j(1 + \beta)}{\alpha^{j+2}}]$$

$$- \sum_{i=0}^{j-1} [\frac{(j - i)(j - i - 1)}{\alpha^{j-i+2}} + \frac{(j - i)(1 + \beta)}{\alpha^{j-i+2}}] \binom{n}{i} \beta^{n-i}\} \mu^2$$

$$\leq \gamma^n \sum_{j=1}^{m} \{\frac{n^2}{\alpha} - \frac{2jn}{\alpha^2} + [\frac{j(j - 1)}{\alpha^3} + \frac{j(1 + \beta)}{\alpha^3}]\} \mu^2,$$

(6.6)

where the negative terms associated with $\sum_{i=0}^{j-1}$ were dropped.

Calculation 3.

$$\sum_{j=1}^{m}(\gamma\alpha)^{j-1}(E_2 - \gamma\beta)^{j*}\{\gamma^n n\}[2\mu(m-j)\lambda](-1)$$

$$= \sum_{j=1}^{m}(\gamma\alpha)^{j-1}\{\gamma^{n-j}[\frac{n}{\alpha^j} - \frac{j}{\alpha^{j+1}}$$

$$+ \sum_{i=0}^{j-1}\binom{n}{i}\beta^{n-i}\alpha^{i-j-1}(j-i)]\}[2\mu(m-j)\lambda](-1)$$

$$\leq \sum_{j=1}^{m}\gamma^n\{\frac{n}{\alpha} - \frac{j}{\alpha^2}\}[2\mu(m-j)\lambda](-1),$$

(6.7)

where the negative terms associated with $\sum_{i=0}^{j-1}$ were dropped;

$$= \sum_{j=1}^{m}\gamma^n\alpha^{-1}\{-2(n\mu)(m\lambda) + j[2n\mu\lambda + \frac{2\mu}{\alpha}(m\lambda)] - j^2(\frac{2\mu\lambda}{\alpha})\}.$$

Calculation 4.

$$\sum_{j=1}^{m}(\gamma\alpha)^{j-1}(E_2 - \gamma\beta)^{j*}\{\gamma^n\}(m-j)^2\lambda^2$$

$$= \sum_{j=1}^{m}(\gamma\alpha)^{j-1}\{\gamma^{n-j}[\frac{1}{\alpha^j} - \frac{1}{\alpha^j}\sum_{i=0}^{j-1}\binom{n}{i}\beta^{n-i}\alpha^i]\}(m-j)^2\lambda^2$$

(6.8)

$$\leq \sum_{j=1}^{m}\gamma^n\alpha^{-1}(m^2 - 2mj + j^2)\lambda^2,$$

where the negative terms associated with $\sum_{i=0}^{j-1}$ were dropped.

Calculation 5. Adding up the right sides of the above three inequalities (6.6), (6.7), and (6.8), and grouping them as a polynomial in j of degree two, we obtain the following:

The term involving $j^0 = 1$ has the factor

$$\mu\frac{1}{\alpha}\sum_{j=1}^{m}[n^2\mu^2 - 2(n\mu)(m\lambda) + (m\lambda)^2] = (m\lambda)(n\mu - m\lambda)^2.$$

The term involving j^2 has the factor

$$\frac{\mu^2}{\alpha^3} - \frac{2\mu\lambda}{\alpha^2} + \frac{\lambda^2}{\alpha} = 0.$$

The term involving j has two parts, one of which has the factor

$$\frac{2n\mu\lambda}{\alpha} + \frac{2\mu m\lambda}{\alpha^2} - \frac{2m\lambda^2}{\alpha} - \frac{2n\mu^2}{\alpha^2} = 0,$$

and the other of which has the factor

$$\mu\sum_{j=1}^{m}(\frac{1+\beta}{\alpha^3} - \frac{1}{\alpha^3})j\mu^2 = (\lambda - \mu)\frac{m(m+1)}{2}\lambda^2.$$

Case 2: (6.2) **holds.** In this case, denote, by $\{h_{m,n} + w_{m,n}\}$, the right side of (6.2), and estimate $\{h_{m,n}\}$. Thanks to (6.5), we derive

$$\{h_{m,n}\} \leq \{K_1\gamma^n(1-\lambda\omega)^{-m}\sum_{k=0}^{n}\beta^{n-k}\alpha^k[2(m-k)+1]\lambda\}$$

$$= \{K_1\gamma^n(1-\lambda\omega)^{-m}[2(m\lambda - n\mu)+\lambda]\},$$

which is less than or equal to

$$\{s_{m,n} + f_{m,n}\},$$

the desired estimate.

Next, to estimate $\{w_{m,n}\}$, the quantity $\delta > 0$ in Case 1 will be used. Owing to

$$r_{m,n} = K_4\mu\rho(|n\mu - m\lambda|),$$

readily seen is

$$\{w_{m,n}\} \leq \{K_4\gamma^n\mu\sum_{j=0}^{n-1}\sum_{i=0}^{j}\beta^{j-i}\alpha^i\rho(|(n\mu - m\lambda)+(i\lambda - j\mu)|)\},$$

which will be written as

$$\{W_{m,n}^{(1)}\} + \{W_{m,n}^{(2)}\}.$$

Here $\{W_{m,n}^{(1)}\}$ is the sum over indices with $|(n\mu - m\lambda)+(i\lambda - j\mu)| < \delta$, while $\{W_{m,n}^{(2)}\}$ is the sum over indices with $|(n\mu - m\lambda)+(i\lambda - j\mu)| \geq \delta$.

An immediate consequence is this estimate

$$\{W_{m,n}^{(1)}\} \leq \{K_4\gamma^n(n\mu)\rho(\delta)\},$$

so we proceed to derive an estimate for $\{W_{m,n}^{(2)}\}$. Due to

$$\frac{|(n\mu - m\lambda)+(i\lambda - j\mu)|^2}{\delta^2} \geq 1,$$

we obtain

$$\{W_{m,n}^{(2)}\} \leq \{K_4\gamma^n\mu\sum_{j=0}^{n-1}\sum_{i=0}^{j}\beta^{j-i}\alpha^i\rho(T)\frac{[(n\mu - m\lambda)+(i\lambda - j\mu)]^2}{\delta^2}\}.$$

This will yield

$$\{W_{m,n}^{(2)}\} \leq \{K_4\gamma^n\frac{\rho(T)}{\delta^2}[(n\mu)(n\mu - m\lambda)^2 + (\lambda - \mu)\frac{n(n-1)}{2}\mu]\},$$

because of the calculations:

$$[(n\mu - m\lambda)+(i\lambda - j\mu)]^2$$

$$= (n\mu - m\lambda)^2 + 2(n\mu - m\lambda)(i\lambda - j\mu)+(i\lambda - j\mu)^2;$$

$$(i\lambda - j\mu)^2 = j^2\mu^2 - 2(j\mu\lambda)i + \lambda^2 i^2;$$

$$\mu\sum_{j=0}^{n-1}\sum_{i=0}^{j}\beta^{j-i}\alpha^i(n\mu - m\lambda)^2 = (n\mu)(n\mu - m\lambda)^2;$$

$$\sum_{i=0}^{j}\beta^{j-i}\alpha^i(i\lambda - j\mu) = (\alpha j)\lambda - j\mu = 0;$$

$$\mu \sum_{j=0}^{n-1} \sum_{i=0}^{j} \beta^{j-i} \alpha^i (i\lambda - j\mu)^2$$

$$= \mu \sum_{j=0}^{n-1} [j^2\mu^2 - 2(j\mu\lambda)(\alpha j) + \lambda^2[\alpha^2 j(j-1) + (\alpha j)]]$$

$$= \mu \sum_{j=0}^{n-1} [j^2(\mu - \alpha\lambda)^2 + j(\lambda^2\alpha)(1-\alpha)]$$

$$= \mu(\lambda - \mu) \sum_{j=0}^{n-1} j = (\lambda - \mu)\frac{n(n-1)}{2}\mu^2.$$

Since $\lambda > \mu$ and $m \geq n$, it follows that

$$\{w_{m,n}\} = \{W_{m,n}^{(1)}\} + \{W_{m,n}^{(2)}\} \leq \{d_{m,n}\}.$$

The proof is complete. $\square$

CHAPTER 3

Linear Autonomous Parabolic Equations

1. Introduction

In this chapter, linear autonomous, parabolic initial-boundary value problems will be solved by utilizing the results in Chapter 1. The obtained solutions will be limit or strong ones under ordinary assumptions. But under stronger assumptions, they will be classical solutions. It is this case that will be demonstrated here. To see how a strong solution is derived, the reader is referred to Chapters 4, 6, and 9.

Let be considered the linear, autonomous, parabolic, initial-boundary value problem with the Robin boundary condition

$$u_t(x,t) = a(x)u_{xx}(x,t) + b(x)u_x(x,t) + c(x)u(x,t),$$
$$(x,t) \in (0,1) \times (0,\infty);$$
$$u_x(0,t) = \beta_0 u(0,t), \quad u_x(1,t) = -\beta_1 u(1,t); \tag{1.1}$$
$$u(x,0) = u_0(x).$$

Here β_0 and β_1 are two positive constants, and $a(x), b(x)$, and $c(x)$ are real-valued, continuous functions on $[0,1]$. Additional restrictions are that, for all x, $c(x)$ is non-positive, and that, for all x, $a(x)$ is greater than or equal to some positive constant δ_0.

(1.1) will be written as the linear Cauchy problem

$$\frac{d}{dt}u(t) = Fu(t), \quad t > 0;$$
$$u(0) = u_0, \tag{1.2}$$

where the linear operator

$$F : D(F) \subset C[0,1] \longrightarrow C[0,1]$$

is defined by

$$Fv = a(x)v'' + b(x)v' + c(x)v;$$
$$v \in D(F) \equiv \{w \in C^2[0,1] : w'(j) = (-1)^j \beta_j w(j), \quad j = 0, 1\}.$$

It will be shown by using the theory in Chapter 1 that the equation (1.2) and then the equation (1.1), for $u_0 \in D(F)$ with $Fu_0 \in \overline{D(F)}$, have a unique classical solution given by

$$u(t) = \lim_{n \to \infty} \left(I - \frac{t}{n}F\right)^{-n} u_0 = \lim_{\nu \to 0} (I - \nu F)^{-[\frac{t}{\nu}]} u_0.$$

The same result holds true for (1.1) with the Robin boundary condition replaced by the Dirichlet or the Neumann or the periodic one:

- $u(0,t) = 0 = u(1,t)$ (Dirichlet condition).
- $u_x(0,t) = 0 = u_x(1,t)$ (Neumann condition) [15].

- $u(0,t) = u(1,t), \quad u_x(0,t) = u_x(1,t)$ (Periodic condition) [15].

In addition to (1.1), higher space dimensional analogue of (1.1) will be the other subject to be studied in this chapter. It takes the form

$$u_t(x,t) = \sum_{i,j=1}^{N} a_{ij}(x)D_{ij}u(x,t) + \sum_{i}^{N} b_i(x)D_iu(x,t)$$

$$+ c(x)u(x,t), \quad (x,t) \in \Omega \times (0,\infty); \tag{1.3}$$

$$\frac{\partial}{\partial \hat{n}}u(x,t) + \beta_2(x)u(x,t) = 0, \quad x \in \partial\Omega;$$

$$u(x,0) = u_0(x);$$

where Ω is a bounded, smooth domain in $\mathbb{R}^N$, and $N \geq 2$ is a positive integer; $x = (x_1, x_2, \ldots, x_N)$, and $D_{ij}u = \sum_{i,j=1}^{N} \frac{\partial^2}{\partial x_i \partial x_j}u$; $u_t = \frac{\partial}{\partial t}u$ and $D_iu = \frac{\partial}{\partial x_i}u$; $\partial\Omega$ is the boundary of Ω, and $\frac{\partial}{\partial \hat{n}}u$ is the outer normal derivative of u; $\beta_2(x)$, for all x, is greater than or equal to some positive constant δ_0.

Additional assumptions are:

- $0 < \mu < 1$.
- $a_{ij}(x), b_i(x)$, and $c(x)$ are μ-Holder continuous functions on $\overline{\Omega}$.
- $\beta_2(x)$ is continuously differentiable on $\overline{\Omega}$, with its first partial derivatives μ-Holder continuous.
- $\delta_0|\xi|^2 \leq \sum_{i,j=1}^{N} a_{ij}(x)\xi_i\xi_j \leq \frac{1}{\delta_0}|\xi|^2$ for all x, ξ.
- $c(x) \leq 0$ for all x.

The corresponding linear Cauchy problem will be

$$\frac{d}{dt}u(t) = Gu(t), \quad t > 0;$$

$$u(0) = u_0, \tag{1.4}$$

in which the linear operator

$$G : D(G) \subset C(\overline{\Omega}) \longrightarrow C(\overline{\Omega})$$

is defined by

$$Gv = \sum_{i,j=1}^{N} a_{ij}(x)D_{ij}v + \sum_{i=1}^{N} b_i(x)D_iv + c(x)v;$$

$$v \in D(G) \equiv \{w \in C^{2+\mu}(\overline{\Omega}) : \frac{\partial}{\partial \hat{n}}w + \beta_2(x)w = 0 \quad \text{on } \partial\Omega\}.$$

The theory in Chapter 1 will be employed again to prove that, for $u_0 \in D^2(G)$, the quantity

$$u(t) = \lim_{n\to\infty} (I - \frac{t}{n}G)^{-n}u_0$$

$$= \lim_{\nu\to 0} (I - \nu G)^{-[\frac{t}{\nu}]}u_0$$

is the unique classical solution for equation (1.4) or, equivalently, equation (1.3).

The same result holds true for (1.3) with the Robin boundary condition replaced by the Dirichlet or the Neumann one:

- $u(x,t) = 0$ on $\partial\Omega$ (Dirichlet condition).
- $\frac{\partial}{\partial \hat{n}}u(x,t) = 0$ on $\partial\Omega$ (Neumann condition) [18].

The rest of this chapter is organized as follows. Section 2 states the main results, and Sections 3 and 4 prove the main results.

The material of this chapter is based on [25].

2. Main Results

THEOREM 2.1. *The linear operator F in (1.2) is a closed operator that satisfies both the dissipativity condition (B2) and the range condition (B1) in Chapter 1. As a result, it follows from Theorem 2.1, Chapter 1 that the equation (1.2) and then the equation (1.1) have a unique solution given by*

$$u(t) = \lim_{n \to \infty} (I - \frac{t}{n} F)^{-n} u_0$$
$$= \lim_{\nu \to 0} (I - \nu F)^{-[\frac{t}{\nu}]} u_0,$$

if $u_0 \in D(F)$ satisfies $F u_0 = [a(x) u_0'' + b(x) u_0' + c(x) u_0] \in \overline{D(F)}$. More smoothness of $u(t)$ in t follows from Theorem 2.2, Chapter 1, if we further restrict u_0.

The same results hold true for the equation (1.1) with the Robin boundary condition replaced by the Dirichlet or the Neumann or the periodic one.

Remark.

- In order for $F u_0$ to be in $\overline{D(F)}$, more smoothness assumptions should be imposed on the coefficient functions $a(x), b(x)$, and $c(x)$.
- The condition $c(x) \le 0$ can be replaced by the condition $c(x) \le \omega$ for some positive constant ω. This is because, in that case, the operator $(F - \omega I)$ is instead considered.

THEOREM 2.2. *The linear operator G in (1.4) satisfies both the dissipativity condition (B2) and the weaker range condition $(B1)'$ in Chapter 1, and hence Theorem 2.3 in Chapter 1 implies the existence of the quantity*

$$u(t) = \lim_{n \to \infty} (I - \frac{t}{n} G)^{-n} u_0$$
$$= \lim_{\nu \to 0} (I - \nu G)^{-[\frac{t}{\nu}]} u_0$$

for $u_0 \in D(G)$. In fact, this $u(t)$, for $u_0 \in D^2(G)$, is a unique classical solution of equation (1.4) or, equivalently, equation (1.3).

The same results are valid for the equation (1.3) with the Robin boundary condition replaced by the Dirichlet or the Neumann one.

Remark.

- The condition $u_0 \in D^2(G)$ requires more smoothness assumptions on the coefficient functions $a_{ij}(x), b_i(x)$, and $c(x)$.
- The condition $c(x) \le 0$ can be weakened to $c(x) \le \omega, \omega > 0$, because, in that case, it suffices to consider the operator $(G - \omega I)$.

3. Proof of One Space Dimensional Case

Proof of Theorem 2.1:

PROOF. We now begin the proof, which is composed of three steps.

Step 1. (F satisfies the dissipativity condition (B2).) Let v_1 and v_2 be in $D(F)$, and let $v_1 \neq v_2$ to avoid triviality. By the first and second derivative tests, there result, for some $x_0 \in (0, 1)$,

$$\|v_1 - v_2\|_\infty = |(v_1 - v_2)(x_0)|;$$
$$(v_1 - v_2)'(x_0) = 0;$$
$$(v_1 - v_2)(x_0)(v_1 - v_2)''(x_0) \leq 0.$$

Here $x_0 \in \{0, 1\}$ is impossible in the case of the Robin boundary condition. For, if $x_0 = 0$ and $\|v_1 - v_2\|_\infty = (v_1 - v_2)(0)$, then

$$(v_1 - v_2)'(0) = \beta_0(v_1 - v_2)(0) > 0,$$

and so $(v_1 - v_2)(0)$ cannot be the positive maximum. This contradicts

$$(v_1 - v_2)(0) = \|v_1 - v_2\|_\infty.$$

Other cases can be treated in the same token, where either

$$x_0 = 0 \quad \text{and} \quad (v_1 - v_2)(0) = -\|v_1 - v_2\|_\infty$$

or $x_0 = 1$.

Reasoning simialr to the above applies to the case with the Dirichlet or periodic condition considered. However, $x_0 \in \{0, 1\}$ is possible under the Neumann condition. Nevertheless, it can be handled by the first and second derivative tests.

The dissipativity condition (B2) is then satisfied, as the calculations show:

$$(v_1 - v_2)(x_0)(Fv_1 - Fv_2)(x_0) \leq 0;$$
$$\|v_1 - v_2\|_\infty^2 = (v_1 - v_2)(x_0)(v_1 - v_2)(x_0)$$
$$\leq [(v_1 - v_2)(x_0)]^2 - \lambda(v_1 - v_2)(x_0)(Fv_1 - Fv_2)(x_0)$$
$$\leq \|v_1 - v_2\|_\infty \|(v_1 - v_2) - \lambda(Fv_1 - Fv_2)\|_\infty \quad \text{for all } \lambda > 0.$$

Step 2. From the theory of ordinary differential equations [5], [24, Corollary 2.13, Chapter 4], the range of $(I - \lambda F), \lambda > 0$, equals $C[0, 1]$, so F satisfies the range condition (B1).

Step 3. (F is a closed operator.) Let $v_n \in D(F)$ converge to v, and Fv_n to w. Then, there is a positive constant K, such that

$$\|v_n\|_\infty \leq K;$$
$$\|Fv_n\|_\infty = \|a(x)v_n'' + b(x)v_n' + c(x)v_n\|_\infty \leq K.$$

This, together with the interpolation inequality [1], [13, page 135], implies that $\|v_n'\|_\infty$ and then $\|v_n\|_{C^2[0,1]}$ are uniformly bounded. Hence it follows from the Ascoli-Arzela theorem [33] that a subsequence of v_n' and then itself converge to v'. That v is in $D(F)$ and Fv_n converges to Fv will be true, whence F is closed. For, by uniform covergence theorem [2], if the Robin boundary condition is considered, then

$$v_n'(x) = \int_{y=0}^{x} \frac{Fv_n - b(y)v_n' - c(y)v_n}{a(y)} \, dy + v_n'(0)$$

$$\text{converges to } v'(x) = \int_{y=0}^{x} \frac{w(y) - b(y)v' - c(y)v}{a(y)} \, dy + v'(0);$$

$$v_n'(j) = (-1)^j \beta_j v_n(j) \quad \text{converges to } v'(j) = (-1)^j \beta_j v(j), \quad j = 0, 1;$$

and so
$$v \in D(F) \text{ and } Fv = w$$
by the fundamental theorem of calculus [**2**]. The case with other boundary conditions is treated in the same way.

The proof is complete. □

4. Proof of Higher Space Dimensional Case

Proof of Theorem 2.2:

PROOF. We now begin the proof, which consists of five steps.

Step 1. (G satisfies the dissipativity condition (B2).) Let v_1 and v_2 be in $D(G)$, and let $v_1 \neq v_2$ to avoid triviality. By the first and second derivative tests, there result, for some $x_0 \in \Omega$,

$$\|v_1 - v_2\|_\infty = |(v_1 - v_2)(x_0)|;$$
$$\nabla(v_1 - v_2)(x_0) = 0, \quad \text{(the gradient of } (v_1 - v_2));$$
$$(v_1 - v_2)(x_0)\triangle(v_1 - v_2)(x_0) \leq 0.$$

Here $x_0 \in \partial\Omega$ is impossible in the case of the Robin boundary condition. For, if $x_0 \in \partial\Omega$ and $\|v_1 - v_2\|_\infty = (v_1 - v_2)(x_0)$, then

$$\frac{\partial}{\partial\hat{n}}(v_1 - v_2)(x_0) > 0$$

by the Hopf boundary point lemma [**13**]. But this is a contradiction to

$$\frac{\partial}{\partial\hat{n}}(v_1 - v_2)(x_0) = -\beta_2(x_0)(v_1 - v_2)(x_0) < 0.$$

The case is similar, where $x_0 \in \partial\Omega$ and $\|v_1 - v_2\|_\infty = -(v_1 - v_2)(x_0)$.

Reasoning like the above applies to the case of the Dirichlet boundary condition. However, $x_0 \in \partial\Omega$ is possible in the case of the Neumann boundary condition. Nonetheless, it can be handled again by the first and second derivative tests. This is because the tangential derivative $\frac{\partial}{\partial t}(v_1 - v_2)(x_0)$ is zero by

$$\|v_1 - v_2\|_\infty = (v_1 - v_2)(x_0), \quad x_0 \in \partial\Omega,$$

which, together with $\frac{\partial}{\partial\hat{n}}(v_1 - v_2)(x_0) = 0$, implies the gradient of $(v_1 - v_2)$ at x_0 is zero.

The dissipativity condition (B2) is then satisfied, as the calculations show:

$$(v_1 - v_2)(x_0)(Gv_1 - Gv_2)(x_0) \leq 0;$$
$$\|v_1 - v_2\|_\infty = (v_1 - v_2)(x_0)(v_1 - v_2)(x_0)$$
$$\leq [(v_1 - v_2)(x_0)]^2 - \lambda(v_1 - v_2)(x_0)(Gv_1 - Gv_2)(x_0)$$
$$\leq \|v_1 - v_2\|_\infty \|(v_1 - v_2) - \lambda(Gv_1 - Gv_2)\|_\infty \quad \text{for all } \lambda > 0.$$

Step 2. From the theory of linear, elliptic partial differential equations [**13**], the range of $(I - \lambda G), \lambda > 0$, equals $C^\mu(\overline{\Omega})$, so G satisfies the weaker range condition $(B1)'$ on account of $C^\mu(\overline{\Omega}) \supset D(G)$.

Step 3. It will be shown that $\|u_i\|_{C^{3+\eta}(\overline{\Omega})}, 0 < \eta < 1$, is uniformly bounded if $u_0 \in D(G^2)$, where

$$u_i = (I - \nu G)^{-i}u_0$$

is that in the discretized equation (2.1) in which A is replaced by G.

Let $u_0 \in D(G)$ for a moment. By the dissipativity condition (B2) or using Lemma 4.2 in Chapter 1, we have

$$\|Gu_i\|_\infty = \|\frac{u_i - u_{i-1}}{\nu}\|_\infty \leq \|Gu_0\|_\infty,$$

which, together with relation

$$u_i - u_0 = \sum_{j=1}^{i}(u_j - u_{j-1}),$$

yields a uniform bound for $\|u_i\|_\infty$. Hence, a uniform bound exists for $\|u_i\|_{C^{1+\lambda}(\overline{\Omega})}$ for any $0 < \lambda < 1$, on using the proof of (4.1) in Chapter 5. (Alternatively, it follows that $\|u_i\|_{W^{2,p}(\Omega)}$ is uniformly bounded for any $p > 2$, on using the L^p elliptic estimates [37]. Hence, so is

$$\|u_i\|_{C^{1+\eta}(\overline{\Omega})} = \|(I - \nu G)^{-i}u_0\|_{C^{1+\eta}(\overline{\Omega})}, \quad 0 < \eta < 1,$$

as a result of the Sobolev embedding theorem [1, 13].) This, applied to the relation

$$Gu_i = (I - \nu G)^{-i}(Gu_0),$$

shows the same thing for $\|Gu_i\|_{C^{1+\eta}(\overline{\Omega})}$, if $Gu_0 \in D(G)$, that is, if $u_0 \in D(G^2)$. Therefore, $\|u_i\|_{C^{3+\eta}(\overline{\Omega})}$ is uniformly bounded if $u_0 \in D(G^2)$, on employing the Schauder global estimates [13].

Step 4. (Existence of a solution) The result in Step 3, together with the Ascoli-Arzela theorem [33], implies that, on putting $i = [\frac{t}{\nu}]$, a subsequence of u_i and then itself, converge to $u(t)$ as $\nu \longrightarrow 0$, with respect to the topology in $C^{3+\lambda}(\overline{\Omega})$ for any $0 < \lambda < 1$. Consequently, as in (6.4), (6.5), and (6.6) in Section 6 of Chapter 1, we have eventually

$$\frac{du(t)}{dt} = Bu(t) = \lim_{n\to\infty}(I - \frac{t}{n}G)^{-n}(Gu_0)$$
$$u(0) = u_0.$$

Thus $u(t)$ is a solution.

Step 5. (Uniqueness of a solution) Let $v(t)$ be another solution. Then, by the first and second derivative tests, we have, for $x_0 \in \Omega$,

$$\|u(t) - v(t)\|_\infty = |[u(t) - v(t)](x_0)|;$$
$$\nabla[u(t) - v(t)](x_0) = 0;$$
$$[u(t) - v(t)](x_0)\triangle[u(t) - v(t)](x_0) \leq 0.$$

Thus it follows that

$$\frac{d}{dt}\|u(t) - v(t)\|_\infty^2 = \frac{d}{dt}[u(t) - v(t)]^2(x_0)$$

$$= 2[u(t) - v(t)](x_0)\frac{d}{dt}[u(t) - v(t)](x_0)$$
$$= 2[u(t) - v(t)](x_0)[Gu(t) - Gv(t)](x_0)$$
$$\leq 0.$$

This implies

$$\|u(t) - v(t)\|_\infty \leq \|u(0) - v(0)\|_\infty$$
$$= \|u_0 - u_0\|_\infty = 0,$$

from which uniqueness of a solution results.

The proof is complete.   □

CHAPTER 4

Nonlinear Autonomous Parabolic Equations

1. Introduction

In this chapter, nonlinear autonomous, parabolic initial-boundary value problems will be solved by making use of the results in Chapter 1. The obtained solutions will be strong ones under suitable assumptions.

Thus the linear cases in Chapter 3 will be extended to the nonlinear cases here.

Let be considered first the case with space dimension equal to one, which is this nonlinear equation with the nonlinear Robin boundary condition

$$
\begin{aligned}
u_t(x,t) &= \alpha(x, u_x)u_{xx}(x,t) + g(x, u, u_x), \\
&\quad (x,t) \in (0,1) \times (0, \infty); \\
u_x(0,t) &\in \beta_0(u(0,t)), \quad u_x(1,t) \in -\beta_1(u(1,t)); \\
u(x,0) &= u_0(x).
\end{aligned}
\tag{1.1}
$$

Here made are the assumptions:

- $\beta_0, \beta_1 : \mathbb{R} \longrightarrow \mathbb{R}$, are multi-valued, maximal monotone functions with $0 \in \beta_0(0) \cap \beta_1(0)$.
- $\alpha(x,p)$ and $g(x,z,p)$ are real-valued, continuous functions of their arguments $x \in [0,1]$, $z \in \mathbb{R}$, and $p \in \mathbb{R}$.
- $\alpha(x,p)$ is greater than or equal to some positive constant δ_0 for all $x \in [0,1]$ and $p \in \mathbb{R}$.
- $g(x,z,p)$ is monotone non-increasing in z for each x and p; that is,

$$
(z_2 - z_1)[g(x, z_2, p) - g(x, z_1, p)] \leq 0.
$$

- $g(x,z,p)/\alpha(x,p)$ is of at most linear growth in p, that is, for some positive, continuous function $M_0(x,z)$,

$$
|g(x,z,p)/\alpha(x,p)| \leq M_0(x,z)(1 + |p|).
$$

(1.1) will be written as the nonlinear Cauchy problem

$$
\begin{aligned}
\frac{d}{dt}u(t) &= Hu(t), \quad t > 0; \\
u(0) &= u_0,
\end{aligned}
\tag{1.2}
$$

where the nonlinear, multi-valued operator

$$
H : D(H) \subset C[0,1] \longrightarrow C[0,1]
$$

is defined by

$$
\begin{aligned}
Hv &= \alpha(x, v')v'' + g(x, v, v'); \\
v &\in D(H) \equiv \{w \in C^2[0,1] : w'(j) \in (-1)^j \beta_j w(j), j = 0, 1\}.
\end{aligned}
$$

It will be shown by using the theory in Chapter 1 that the equation (1.2) and then the equation (1.1), for $u_0 \in D(H)$, have a strong solution given by

$$u(t) = \lim_{n\to\infty} (I - \frac{t}{n}H)^{-n} u_0$$
$$= \lim_{\nu\to 0} (I - \nu H)^{-[\frac{t}{\nu}]} u_0.$$

The other case to be studied is with space dimension greater than one, which takes the form

$$u_t(x,t) = \alpha_0(x, Du)\triangle u(x,t) + g_0(x, u, Du),$$
$$(x,t) \in \Omega \times (0,\infty);$$

$$\frac{\partial}{\partial \hat{n}} u(x,t) + \beta(x,u) = 0, \quad x \in \partial\Omega;$$

$$u(x,0) = u_0(x).$$

(1.3)

Here

- Ω is a bounded, smooth domain in $\mathbb{R}^N$, and $N \geq 2$ is a positive integer;
- $x = (x_1, x_2, \ldots, x_N)$, and $\triangle u = \sum_{i=1}^{N} \frac{\partial^2}{\partial x_i^2} u$;
- $u_t = \frac{\partial}{\partial t} u$, $D_i u = \frac{\partial}{\partial x_i} u$, and $Du = (D_1 u, D_2 u, \ldots, D_N u)$;
- $\partial\Omega$ is the boundary of Ω, and $\frac{\partial}{\partial \hat{n}} u$ is the outer normal derivative of u;

and additional assumptions are:

- $\alpha_0(x,p)$ and $g_0(x,z,p)$ are real-valued, continuously differentiable functions of their arguments $x \in \overline{\Omega}, z \in \mathbb{R}$, and $p \in \mathbb{R}^N$, and their first partial derivatives are μ-Holder continuous, where $0 < \mu < 1$.
- $\beta(x,z)$ is twice continuously differentiable function of their arguments $x \in \overline{\Omega}$ and $z \in \mathbb{R}$, and its second partial derivatives are μ-Holder continuous.
- $\alpha_0(x,p)$ is greater than or equal to some positive constant δ_1 for all $x \in [0,1]$ and $p \in \mathbb{R}$.
- $g_0(x,z,p)/\alpha_0(x,p)$ is of at most linear growth in p, that is, for some positive, continuous function $M_1(x,z)$,

$$|g_0(x,z,p)/\alpha_0(x,p)| \leq M_1(x,z)(1+|p|).$$

- $g_0(x,z,p)$ is monotone non-increasing in z, that is,

$$(z_2 - z_1)[g_0(x,z_2,p) - g_0(x,z_1,p)] \leq 0.$$

- $\beta(x,z)$ is strictly monotone increasing in z, that is, for some positive constant δ_1,

$$\beta_z(x,z) \geq \delta_1.$$

The corresponding nonlinear Cauchy problem will be

$$\frac{d}{dt} u(t) = Ju(t), \quad t > 0;$$
$$u(0) = u_0,$$

(1.4)

in which the nonlinear operator

$$J : D(J) \subset C(\overline{\Omega}) \longrightarrow C(\overline{\Omega})$$

is defined by

$$Jv = \alpha_0(x, Dv)\triangle v + g_0(x, v, Dv);$$

$$v \in D(J) \equiv \{w \in C^{2+\mu}(\overline{\Omega}) : \frac{\partial}{\partial \hat{n}} w + \beta(x, w) = 0 \quad \text{on } \partial\Omega\}.$$

The theory in Chapter 1 will be employed again to prove that, for $u_0 \in D(G)$, the quantity

$$u(t) = \lim_{n \to \infty} (I - \frac{t}{n}G)^{-n} u_0$$
$$= \lim_{\nu \to 0} (I - \nu G)^{-[\frac{t}{\nu}]} u_0$$

is a strong solution for equation (1.4) or, equivalently, equation (1.3).

The rest of this chapter is organized as follows. Section 2 states the main results, and Sections 3 and 4 prove the main results.

2. Main Results

THEOREM 2.1. *The nonlinear operator H in (1.2) satisfies both the dissipativity condition (A2) and the range condition (A1) in Chapter 1. As a result, it follows from Theorem 2.4, Chapter 1 that the quantity*

$$u(t) = \lim_{n \to \infty} (I - \frac{t}{n}H)^{-n} u_0$$
$$= \lim_{\nu \to 0} (I - \nu H)^{-[\frac{t}{\nu}]} u_0$$

exists for $u_0 \in \overline{D(H)}$.

Furthermore, the H also satisfies the embedding condition (A3) in Chapter 1, of embeddedly quasi-demi-closedness, and hence, by Theorem 2.5 in Chapter 1, the $u(t)$, for $u_0 \in D(H)$, is not only a limit solution but also a strong one for the equation (1.2) and then the equation (1.1). The $u(t)$ will also satisfy the middle equation in (1.1).

Theorem 2.1 is taken from [20].

Remark.

- The condition

$$(z_2 - z_1)[g(x, z_2, p) - g(x, z_1, p)] \le 0$$

can be replaced by the condition

$$(z_2 - z_1)[g(x, z_2, p) - g(x, z_1, p)] \le \omega(z_2 - z_1)^2$$

for some positive constant ω. This is because, in that case, the operator $(H - \omega I)$ is instead considered.

THEOREM 2.2. *The nonlinear operator J in (1.4) satisfies the three conditions in Chapter 1: the dissipativity condition (A2), the weaker range condition (A1)', and the embedding condition (A3) of embeddedly quasi-demi-closedness. Consequently, Theorem 2.6 in Chapter 1 implies that the quantity*

$$u(t) = \lim_{n \to \infty} (I - \frac{t}{n}J)^{-n} u_0$$
$$= \lim_{\nu \to 0} (I - \nu J)^{-[\frac{t}{\nu}]} u_0$$

exists for $u_0 \in \overline{D(G)}$, and that this $u(t)$, for $u_0 \in D(G)$, is not only a limit solution but also a strong one for the equation (1.4) or, equivalently, equation (1.3). The $u(t)$ will also satisfy the middle equation in (1.3).

Theorem 2.2 is taken from [19].

Remark.

- The condition

$$(z_2 - z_1)[g_0(x, z_2, p) - g_0(x, z_1, p)] \leq 0$$

can be weakened to

$$(z_2 - z_1)[g_0(x, z_2, p) - g_0(x, z_1, p)] \leq \omega(z_2 - z_1)^2$$

for some $\omega > 0$, because, in that case, it suffices to consider the operator $(J - \omega I)$.

3. Proof of One Space Dimensional Case

Proof of Theorem 2.1:

PROOF. We now begin the proof, which is composed of five steps.

Step 1. (H satisfies the dissipativity condition (A2).) This is done as in solving Example 3.3, Chapter 1.

Let v_1 and v_2 be in $D(H)$, and let $v_1 \neq v_2$ to avoid triviality. By the first and second derivative tests, there result, for some $x_0 \in [0, 1]$,

$$\|v_1 - v_2\|_\infty = |(v_1 - v_2)(x_0)|;$$
$$(v_1 - v_2)'(x_0) = 0;$$
$$(v_1 - v_2)(x_0)(v_1 - v_2)''(x_0) \leq 0.$$

Here $x_0 \in \{0, 1\}$ is possible, due to the boundary conditions in $D(H)$. For, if $x_0 = 0$ and $\|v_1 - v_2\|_\infty = (v_1 - v_2)(0)$, then the monotonicity of β_0 and the positivity of $(v_1 - v_2)(0)$ implies

$$(v_1 - v_2)'(0) \geq 0.$$

From this, there must $(v_1 - v_2)'(0) = 0$ because if $(v_1 - v_2)'(0) > $ occurs, then $(v_1 - v_2)(0)$ cannot be the positive maximum. Other cases can be treated similarly.

The dissipativity condition (A2) is then satisfied, as the calculations show:

$$(v_1 - v_2)(x_0)(Hv_1 - Hv_2)(x_0)$$
$$= (v_1 - v_2)(x_0)[\alpha(x_0, v_1')(v_1 - v_2)''(x_0)$$
$$+ g(x_0, v_1, v_1') - g(x_0, v_2, v_1')] \leq 0;$$
$$\|v_1 - v_2\|_\infty^2 = (v_1 - v_2)(x_0)(v_1 - v_2)(x_0)$$
$$\leq [(v_1 - v_2)(x_0)]^2 - \lambda(v_1 - v_2)(x_0)(Hv_1 - Hv_2)(x_0)$$
$$\leq \|v_1 - v_2\|_\infty \|(v_1 - v_2) - \lambda(Hv_1 - Hv_2)\|_\infty \quad \text{for all } \lambda > 0.$$

Step 2. The proof of Example 3.1 in Chapter 1 proved that, for each $h \in C[0, 1]$ and for all $\lambda = \mu^2 > 0$ where $0 < \mu < [\log(3)]^{-1}$, the equation

$$v - \lambda v'' = h,$$
$$v'(j) \in (-1)^j \beta_j(v(j)), \quad j = 0, 1$$

has a unique solution. Hence, the operator

$$(I - \lambda E)^{-1} : C[0, 1] \longrightarrow C^1[0, 1]$$

exists, where the operator $E : D(E) \subset C[0, 1] \longrightarrow C[0, 1]$ is defined by

$$Ev = v'';$$

$$D(E) \equiv \{w \in C^2[0,1] : w'(j) \in (-1)^j \beta_j(u(j)), j = 0, 1\}.$$

It is also continuous because, for

$$h_1, h_2 \in C[0,1];$$
$$v_j = (I - \lambda E)^{-1} h_j, j = 1, 2,$$

there results $\|v_2 - v_1\|_\infty \leq \|h_2 - h_1\|_\infty$ by the dissipativity of E, whence

$$\|v_2'' - v_1''\|_\infty \leq 2\|h_2 - h_1\|_\infty/\lambda;$$
$$\|v_2' - v_1'\|_\infty \leq K_0\|h_2 - h_1\|_\infty \quad \text{for some constant } K_0 > 0.$$

Here the interpolation inequality was used [1], [13, page 135]: for each $\epsilon > 0$, there is a positive $C(\epsilon)$ such that, for $v \in C^2[0,1]$,

$$\|v'\|_\infty \leq \epsilon\|v''\|_\infty + C(\epsilon)\|v\|_\infty.$$

The above also shows that $(I - \lambda E)^{-1}$ is a compact operator, where the Ascoli-Arzela theorem [33] was used.

Step 3. (H satisfies the range condition (A1).) Let $h \in C[0,1]$ be given, and let $\lambda = \mu^2 > 0$ where $0 < \mu < [\log(3)]^{-1}$. It will be shown that the equation

$$u - \lambda[\alpha(x, u, u')u'' + g(x, u, u')] = h \qquad (3.1)$$
$$u'(j) \in (-1)^j \beta_j(u(j)), \quad j = 0, 1$$

has a solution.

Consider the operator equation equation $u = (I - \lambda E)^{-1} W u$, where $(I - \lambda E)^{-1}$ is from Step 2, and the operator $W : C^1[0,1] \longrightarrow C[0,1]$ is defined by

$$Wu = u + \frac{h - u + \lambda g(x, u, u')}{\alpha(x, u')}$$

for $u \in C^1[0,1]$. Solvability of this operator equation will complete the proof.

Truncating W by defining, for each $m \in \mathbb{N}$,

$$W_m u = \begin{cases} Wu, & \text{if } \|u\|_{C^1[0,1]} \leq m; \\ W(\frac{mu}{\|u\|_{C^1[0,1]}}), & \text{if } \|u\|_{C^1[0,1]} > m, \end{cases}$$

it follows that $(I - \lambda E)^{-1} W_m : C^1[0,1] \longrightarrow C^1[0,1]$ is continuous, compact, and uniformly bounded for each m. Hence, the Schauder fixed point theorem [13] implies that

$$(I - \lambda E)^{-1} W_m u_m = u_m$$

for some u_m. The proof is completed if $\|u_{m_0}\|_{C^1} \leq u_{m_0}$ for some m_0, because then, $(I - \lambda E)^{-1} W u_{m_0} = u_{m_0}$.

Assuming, to the contrary, that $\|u_m\|_{C^1} > m$ for all m, we will seek a contradiction. It follows from the definition of W_m that

$$u_m - \lambda u_m'' = v_m + \frac{h - v_m + \lambda g(x, v_m, v_m')}{\alpha(x, v_m')}, \qquad (3.2)$$

where $u_m \in D(E)$ and $v_m = \frac{m u_m}{\|u_m\|_{C^1}}$. Using the relations which are consequences of the first and second derivative tests,

$$\|u_m\|_\infty = |u_m(x_0)|; \quad u_m'(x_0) = 0;$$
$$u_m(x_0)u_m''(x_0) \leq 0$$

for some $x_0 \in [0, 1]$, and multiplying (3.2) by u_m and evaluating it at x_0, it results that

$$0 \le [(1 - \frac{m}{\|u_m\|_{C^1}})u_m^2(x_0) - \lambda u_m(x_0)u_m''(x_0)]\alpha(x_0, 0)$$
$$= (h - v_m)(x_0)u_m(x_0) + \lambda u_m(x_0)g(x_0, v_m, 0)$$
$$\le (h - v_m)(x_0)u_m(x_0).$$

Therefore, $\|v_m\|_\infty \le \|h\|_\infty$. This, combined with (3.2), yields

$$\|v_m''\|_\infty \le \lambda^{-1}K_0(\|h\|_\infty + \|v_m'\|_\infty)$$

for some constant K_0, whence, on employing the above interpolation inequality again [1], [13, page 135], there results $\|v_m\|_{C^2[0,1]} \le K_1$ for constant K_1. But this is a contradiction to $m = \|v_m\|_{C^1} \le \|v_m\|_{C^2}$, if we let $m \longrightarrow \infty$.

Step 4. (H satisfies the embedding condition (A3) of embeddedly quasi-demi-closedness.) [20] Let $v_n \in D(H)$ converge to v in $C[0, 1]$, and let $\|Hv_n\|_\infty$ be uniformly bounded. It will be shown that, for each η in the self-dual space $L^2(0, 1) = (L^2(0, 1))^*$, $\eta(Hv)$ exists and

$$|\eta(Hv_n) - \eta(Hv)| \longrightarrow 0.$$

Here $(C[0, 1]; \|\cdot\|_\infty)$ is continuously embedded into $L^2(0, 1); \|\cdot\|)$.

Since $\|v_n\|_\infty$ and $\|Hv_n\|_\infty$ are uniformly bounded, so is $\|v_n\|_{C^2[0,1]}$ by the interpolation inequality [1], [13, page 135]. Hence, by the Ascoli-Arzela theorem [33], a subsequence of v_n and then itself converge in $C^1[0, 1]$ to v. Also, v_n is uniformly bounded in the Hilbert space $W^{2,2}(0, 1)$, whence, by Alaoglu theorem [36], a subsequence of v_n and then itself converge weakly to v [36]. It follows that, for each $\eta \in L^2(0, 1)$,

$$|\eta(Hv_n) - \eta(Hv)| = |\int_0^1 (Hv_n - Hv)\eta \, dx| \longrightarrow 0$$

because

$$\int_0^1 \alpha(x, v')(v'' - v_n'')\eta \, dx + \int_0^1 [\alpha(x, v') - \alpha(x, v_n')]v_n''\eta \, dx$$
$$+ \int_0^1 [g(x, v, v') - g(x, v_n, v_n')]\eta \, dx$$
$$\longrightarrow 0.$$

Therefore H satisfies the embedding condition (A3).

Step 5. ($u(t)$ satisfies the middle equation in (1.1).) Consider the discretized equation

$$u_i - \nu H u_i = u_{i-1},$$
$$u_i \in D(H),$$

where $i = 1, 2, \ldots$, $\nu > 0$ satisfies $\nu < \lambda_0$ for which $\nu\omega < 1$, and

$$u_i = (I - \nu H)^{-i}u_0$$

exists uniquely by the range condition (A1) and the dissipativity condition (A2). On putting $i = [\frac{t}{\nu}]$, it follows that

$$\lim_{\nu \to 0} u_i = \lim_{\nu \to 0}(I - \nu H)^{-[\frac{t}{\nu}]}u_0 = u(t).$$

On the other hand, by utilizing the dissipativity condition (A2), we have

$$\|Hu_i\|_\infty = \|\frac{u_i - u_{i-1}}{\nu}\|_\infty \leq \|Hu_0\|_\infty.$$

This, combined with the relation

$$u_i - u_0 = \sum_{j=1}^{i}(u_j - u_{j-1}),$$

yields a bound for $\|u_i\|_\infty$. Those, in turn, result in a bound for $\|u_i\|_{C^2[0,1]}$ by the interpolation inequality [1], [13, page 135]. Therefore it follows from Ascoli-Arzela theorem [33] that a subsequence of u_i and then itself converge to a limit in $C^1[0,1]$, as $\nu \longrightarrow 0$. This limit equals $u(t)$ as shown above. Consequently, $u(t)$ satisfies the middle equation in (1.1), as u_i does so.

The proof is complete. □

4. Proof of Higher Space Dimensional Case

Proof of Theorem 2.2:

PROOF. We now begin the proof, which consists of four steps.

Step 1. (J satisfies the dissipativity condition (A2).) This will be proved as in solving Example 3.2 of Chapter 1. Let v_1 and v_2 be in $D(J)$, and let $v_1 \neq v_2$ to avoid triviality. By the first and second derivative tests, there result, for some $x_0 \in \Omega$,

$$\|v_1 - v_2\|_\infty = |(v_1 - v_2)(x_0)|;$$
$$D(v_1 - v_2)(x_0) = 0, \quad \text{(the gradient of } (v_1 - v_2));$$
$$(v_1 - v_2)(x_0)\triangle(v_1 - v_2)(x_0) \leq 0.$$

Here $x_0 \in \partial\Omega$ is impossible, due to the boundary condition in $D(J)$. For, if $x_0 \in \partial\Omega$ and $\|v_1 - v_2\|_\infty = (v_1 - v_2)(x_0)$, then

$$\frac{\partial}{\partial\hat{n}}(v_1 - v_2)(x_0) > 0$$

by the Hopf boundary point lemma [13]. But this is a contradiction to

$$\frac{\partial}{\partial\hat{n}}(v_1 - v_2)(x_0) = -\beta_2(x_0)(v_1 - v_2)(x_0) < 0.$$

The case is similar, where $x_0 \in \partial\Omega$ and $\|v_1 - v_2\|_\infty = -(v_1 - v_2)(x_0)$.

The dissipativity condition (A2) is then satisfied, as the calculations show:

$$(v_1 - v_2)(x_0)(Jv_1 - Jv_2)(x_0)$$
$$= (v_1 - v_2)(x_0)[\alpha_0(x_0, Dv_1)\triangle(v_1 - v_2)(x_0)$$
$$+ g_0(x_0, v_1, Dv_1) - g_0(x_0, v_2, Dv_1)] \leq 0;$$
$$\|v_1 - v_2\|_\infty = (v_1 - v_2)(x_0)(v_1 - v_2)(x_0)$$
$$\leq [(v_1 - v_2)(x_0)]^2 - \lambda(v_1 - v_2)(x_0)(Jv_1 - Jv_2)(x_0)$$
$$\leq \|v_1 - v_2\|_\infty\|(v_1 - v_2) - \lambda(Jv_1 - Jv_2)\|_\infty \quad \text{for all } \lambda > 0.$$

Step 2. (J satisfies the weaker range condition $(A1)'$.) Let $h \in C^\mu(\overline{\Omega})$ and $\lambda > 0$. It will be shown that the equation

$$u - \lambda[\alpha_0(x, Du)\triangle u + g_0(x, u, Du)] = h \quad \text{in } \Omega;$$
$$\frac{\partial u}{\partial \hat{n}} + \beta(x, u) = 0 \quad \text{on } \partial\Omega, \tag{4.1}$$

has a solution, so that the range of $(I - \lambda J)$ contains $C^\mu(\overline{\Omega}) \supset D(J)$.

To this end, use the method of continuity [**13**] by considering the family of equations, indexed by $t \in [0, 1]$:

$$L_t u = tLu + (1-t)L_0 u = h \quad \text{in } \Omega;$$
$$N_t u = t[\frac{\partial u}{\partial \hat{n}} + \beta(x, u)] + (1-t)(\frac{\partial u}{\partial \hat{n}} + u) = 0 \quad \text{on } \partial\Omega;$$

where

$$L_0 u = u - \triangle u;$$
$$Lu = u - [\alpha_0(x, Du)\triangle u + g_0(x, u, Du)].$$

Here note that $\lambda = 1$ was assumed, which is sufficient because $\lambda\alpha_0$ and λg_0 have the same properties as α_0 and g_0 have. By defining the set

$$S = \{t \in [0, 1] : L_t u = h \quad \text{in } \Omega \text{ and } N_t u = 0 \text{ on } \partial\Omega \text{ for some } u \in C^{2+\mu}(\overline{\Omega})\},$$

it will be shown that S is open, closed, and not empty in $[0, 1]$. Hence $S = [0, 1]$ and so, $1 \in S$, proving that (4.1) is solved.

It is readily seen from linear elliptic theory [**13**] that $0 \in S$, so we proceed to verify that S is both open and closed.

To show that S is open, let

$$B_1 = C^{2+\mu}(\overline{\Omega});$$
$$B_2 = C^\mu(\overline{\Omega}) \times C^{1+\mu}(\partial\Omega),$$

and define the nonlinear operator

$$G : D(G) \equiv B_1 \times [0, 1] \longrightarrow B_2;$$
$$G(u, t) = (L_t u - h, N_t u).$$

By assuming $t_0 \in S$, that is, assuming $G(u_0, t_0) = 0$ for some $u_0 \in C^{2+\mu}(\overline{\Omega})$, an open neighborhood of t_0 in S will exist as a consequence of the implicit function theorem [**13**]. Indeed, because G is continuously Frechet differentiable, the partial Frechet derivative G_{u_0} at $(u_0, t_0) : B_1 \longrightarrow B_2$ exists and is given by

$$G_{u_0}(v) = ([-t_0\alpha_0 - (1-t_0)]\triangle v - t_0[(g_0)_{p_i} + (\alpha_0)_{p_i}\triangle u_0]D_i v$$
$$+ [t_0(1 - g_z) + (1-t_0)]v, \frac{\partial v}{\partial \hat{n}} + [t_0\beta_z + (1-t_0)]v)$$

for $v \in B_1$. Here

$$\alpha_0 = \alpha_0(x, Du_0); \quad (\alpha_0)_{p_i} = (\alpha_0)_{p_i}(x, Du_0);$$
$$(g_0)_z = (g_0)_z(x, u_0, Du_0); \quad (g_0)_{p_i} = (g_0)_{p_i}(x, u_0, Du_0);$$
$$\beta_z = \beta_z(x, u_0).$$

Now that G_{u_0} is an invertible, linear operator by linear elliptic theory [**13**], where β_z lies in $C^{1+\mu}(\overline{\Omega})$, the implicit function theorem [**13**] implies the existence of an open neighborhood of t_0 in S.

That S is a closed set will be a result of some estimates. Let t_k be a sequence in S that converges to t_0, and we will show $t_0 \in S$. By the definition of S, there exists a sequence $u_{t_k} \in B_1$ such that

$$L_{t_k} u_{t_k} = h \quad \text{in } \Omega;$$
$$N_{t_k} u_{t_k} = 0 \quad \text{on } \partial\Omega, \tag{4.2}$$

whence, owing to the mean value theorem,

$$u_{t_k} - [t_k \alpha_0 + (1 - t_k)] \Delta u_{t_k} - t_k g_0 = h \quad \text{in } \Omega;$$
$$[t_k \int_0^1 \beta_z \, d\theta + (1 - t_k)] u_{t_k} + \frac{\partial u_{t_k}}{\partial \hat{n}} = -t_k \beta(x, 0) \quad \text{on } \partial\Omega. \tag{4.3}$$

Here the arguments of α_0 and g_0 were suppressed, and the bottom bracket will be, for convenience, written as $[\gamma_{t_k}(x) u_{t_k} + \frac{\partial u_{t_k}}{\partial \nu}]$. In order to estimate u_{t_k}, the maximum principle arguments will be used. If the maximum value of $|u_{t_k}|$ occurs at some interior point in Ω, then the first and second derivative tests as in Step 1 will yield

$$\|u_{t_k}\|_\infty \le t_k \|g(x, 0, 0)\|_\infty + \|h\|_\infty.$$

Here taken into account was the monotone non-increasing assumption of $g_0(x, z, p)$ in z.

But if the maximum value occurs on the boundary, then $\|u_{t_k}\|_\infty = \pm u_{t_k}(x_0)$ for some $x_0 \in \partial\Omega$. Considering the case $\|u_{t_k}\|_\infty = u_{t_k}(x_0)$ is sufficient, for which, by the Hopf boundary point lemma [13], $\frac{\partial u_{t_k}(x_0)}{\partial \hat{n}} > 0$. Here $u_{t_k}(x_0) \ne 0$ was assumed to avoid triviality. Since $\gamma_{t_k}(x) \ge \delta_2 > 0$ for some constant $\delta_2 > 0$ by the assumption that $\beta(x, z)$ is stricly monotone increasing in z, there results, from (4.3),

$$\|u_{t_k}\|_\infty = u_{t_k}(x_0) \le [t_k \delta_2 + (1 - t_k)]^{-1} t_k \|\beta(x, 0)\|_\infty.$$

To estimate u_{t_k} further, rewrite (4.3) as

$$\Delta u_{t_k} = \frac{-t_k g_0}{t_k \alpha_0 + (1 - t_k)} + \frac{u_{t_k} - h}{t_k \alpha_0 + (1 - t_k)} \equiv F(x, u_{t_k}, Du_{t_k}) \quad \text{in } \Omega;$$
$$\frac{\partial u_{t_k}}{\partial \nu} = -\gamma_{t_k} u_{t_k} - t_k \beta(x, 0) \equiv H(x, u_{t_k}) \quad \text{on } \partial\Omega, \tag{4.4}$$

from which u_{t_k} has the integral representation [29]

$$u_{t_k} = -\int_{\partial\Omega} Z(x, y) H(y, u_{t_k}) \, d\sigma_y + \int_\Omega Z(x, y) F(y, u_{t_k}, Du_{t_k}) \, dy.$$

Here $Z(x, y)$ is the Green's function of the second kind. Since $[g_0(x, z, p)/\alpha_0(x, p)]$ or $F(x, z, p)$ is of at most linear growth in p and $\|u_{t_k}\|_\infty$ is uniformly bounded, we can differentiate the u_{t_k} for $(1 + \mu)$ times $(0 < \mu < 1)$ in the above integral representation to obtain

$$\|Du_{t_k}\|_{C^\mu(\overline{\Omega})} \le k \|Du_{t_k}\|_\infty + k.$$

Here

$$|D_i Z(x, y)| \le k|x - y|^{1-n};$$
$$|D_{ij} Z(x, y)| \le k|x - y|^{-n};$$

and $\int_\Omega |x - y|^{n(\mu-1)} dx$ is finite for $1 > \mu > 0$ and for bounded y [13, page 159]. Thanks to the interpolation inequality [13, page 135], the estimate

$$\|u_{t_k}\|_{C^{1+\mu}(\overline{\Omega})} \leq k \tag{4.5}$$

is derived. This, combined with the Schauder global estimate [13], yields

$$\|u_{t_k}|_{C^{2+\mu}(\overline{\Omega})} \leq k.$$

Therefore, it follows from the Ascoli-Arzela theorem [33] that a subsequence of u_{t_k} converges to some $u_0 \in C^{2+\gamma}(\overline{\Omega})$, where $0 < \gamma < \mu$. As a result, the equation (4.2) converges to

$$\begin{aligned} L_{t_0} u_0 &= h \quad \text{in } \Omega; \\ N_{t_0} u_0 &= 0 \quad \text{on } \partial\Omega. \end{aligned} \tag{4.6}$$

Now u_0 is not just in $C^{2+\gamma}(\overline{\Omega})$ but also in $C^{2+\mu}(\overline{\Omega})$, from which $t_0 \in S$ and S is closed. Indeed, this is a result of the theory of linear elliptic partial differential equations [13], because u_0 satisfies (4.6) and both the function h and the coefficient functions in (4.6) are in $C^\mu(\overline{\Omega})$.

Step 3. (J satisfies the embedding condition (A3) of embeddedly quasi-demi-closedness.) [20] Let $v_n \in D(G(t_n))$ converge to v in $C(\overline{\Omega})$, and $\|G(t_n)v_n\|_\infty$ be uniformly bounded. It will be shown that, for each η in the self-dual space $L^2(\Omega) = (L^2(\Omega))^*$, $\eta(G(t)v)$ exists and

$$|\eta(G(t_n)v_n) - \eta(G(t)v)| \longrightarrow 0.$$

Here $(C(\overline{\Omega}); \|\cdot\|_\infty)$ is continuously embedded into $L^2(\Omega); \|\cdot\|)$.

Since $\|v_n\|_\infty$ and $\|G(t_n)v_n\|_\infty$ are uniformly bounded, so is $\|v_n\|_{C^{1+\lambda}(\overline{\Omega})}$ for any $0 < \lambda < 1$, using the proof of (4.5). (Alternatively, uniformly bounded are $\|v_n\|_{W^{2,p}(\Omega)}$ for any $p \geq 2$ and then $\|v_n\|_{C^{1+\lambda}(\overline{\Omega})}$ for any $0 < \lambda < 1$, on using the L^p elliptic estimates [37] and the Sobolev embedding theorem [1, 13].) Hence, by the Ascoli-Arzela theorem [33], a subsequence of v_n and then itself converge in $C^1(\overline{\Omega})$ to v. Also, v_n is uniformly bounded in the Hilbert space $W^{2,2}(\Omega)$, whence, by the Alaoglu theorem [36], a subsequence of v_n and then itself converge weakly to v [36]. It follows that, for each $\eta \in L^2(\Omega)$,

$$|\eta(G(t_n)v_n) - \eta(G(t)v)| \longrightarrow 0,$$

because

$$\int_\Omega \alpha_0(x, Dv)(\triangle v - \triangle v_n)\eta \, dx + \int_\Omega [\alpha_0(x, Dv) - \alpha_0(x, Dv_n)]\triangle v_n \eta \, dx$$

$$+ \int_\Omega [g_0(x, v, Dv) - g_0(x, v_n, Dv_n)]\eta \, dx$$

$$\longrightarrow 0.$$

Therefore J satisfies the embedding condition (A3).

Step 4. ($u(t)$ for $u_0 \in D(J)$ satisfies the middle equation in (1.3).) Consider the discretized equation

$$u_i - \nu J u_i = u_{i-1},$$

$$u_i \in D(J)),$$

where $i = 1, 2, \ldots, \nu > 0$ is such that $\nu\omega < 1$, and

$$u_i = (I - \nu J)^{-i} u_0$$

exists uniquely by the weaker range condition $(A1)'$ and the dissipativity condition $(A2)$.

On putting $i = [\frac{t}{\nu}]$, it follows that

$$\lim_{\nu \to 0} u_i = \lim_{\nu \to 0} (I - \nu J)^{-[\frac{t}{\nu}]} u_0 = u(t).$$

On the other hand, by utilizing the dissipativity condition $(A2)$, we have

$$\|Ju_i\|_\infty = \|\frac{u_i - u_{i-1}}{\nu}\|_\infty \le \|Ju_0\|_\infty,$$

which, combined with the relation

$$u_i - u_0 = \sum_{j=1}^{i} (u_j - u_{j-1}),$$

yields a bound for $\|u_i\|_\infty$. It follows that $\|u_i\|_{C^{1+\lambda}(\overline{\Omega})}$ is uniformly bounded for any $0 < \lambda < 1$, on using the proof of (4.5). (Alternatively, those, in turn, result in a bound for $\|u_i\|_{W^{2,p}(\Omega)}$ for any $p \ge 2$, by the L^p elliptic estimates [37]. Hence, a bound exists for $\|u_i\|_{C^{1+\eta}(\overline{\Omega})}$ for any $0 < \eta < 1$, as a result of the Sobolev embedding theorem [1, 13].) Therefore it follows from Ascoli-Arzela theorem [33] that a subsequence of u_i and then itself converge in $C^{1+\mu}(\overline{\Omega})$ to a limit, as $\nu \longrightarrow 0$. This limit equals $u(t)$ as shown above. Consequently, $u(t)$ satisfies the middle equation in (1.3), as u_i does so.

The proof is complete. $\qquad\qquad\qquad\qquad\qquad\qquad\qquad\qquad\qquad\qquad\qquad$ □

CHAPTER 5

Linear Non-autonomous Parabolic Equations

1. Introduction

In this chapter, linear non-autonomous, parabolic initial-boundary value problems will be solved by applying the results in Chapter 2. The obtained solutions will be limit or strong ones under ordinary assumptions. But under stronger assumptions, they will be classical solutions. It is this case that will be shown here. To see how a strong solution is yielded, the reader is referred to Chapters 6 and 9.

Thus the linear autonomous cases in Chapter 3 shall be generalized to the linear non-autonomous cases here.

The case with space dimension equal to one is the linear, non-autonomous equation with the time-dependent Robin boundary condition

$$u_t(x,t) = a(x,t)u_{xx}(x,t) + b(x,t)u_x(x,t) + c(x,t)u(x,t) + f_0(x,t),$$
$$(x,t) \in (0,1) \times (0,T);$$
$$u_x(0,t) = \beta_0(0,t)u(0,t), \quad u_x(1,t) = -\beta_1(1,t)u(1,t);$$
$$u(x,0) = u_0(x).$$

(1.1)

This equation is different from that in Chapter 3, in that the t dependence in the coefficient functions is permitted. Here the six functions

$$a(x,t), b(x,t), c(x,t), f_0(x,t), \beta_0(x,t), \text{ and } \beta_1(x,t)$$

are real-valued and jointly continuous in $x \in [0,1]$ and $t \in [0,T]$ where $T > 0$. Three more restrictions are made. One is that, for all x and t, $c(x,t)$ is non-positive, and the second is that, for all x and t, the functions $a(x,t), \beta_0(x,t)$, and $\beta_1(x,t)$ are greater than or equal to some positive constant δ_0. The third is that the above six functions satisfy, for $x \in [0,1], t, \tau \in [0,T]$, and $\zeta(t)$, a function in t of bounded variation,

$$|a(x,t) - a(x,\tau)|, \quad |b(x,t) - b(x,\tau)| \le |\zeta(t) - \zeta(\tau)|;$$
$$|c(x,t) - c(x,\tau)|, \quad |f_0(x,t) - f_0(x,\tau)| \le |\zeta(t) - \zeta(\tau)|;$$
$$|\beta_j(x,t) - \beta_j(x,\tau)| \le M_0|t - \tau|, \quad j = 0, 1.$$

Here M_0 is a positive constant.

(1.1) will be written as the evolution equation

$$\frac{d}{dt}u(t) = F(t)u(t), \quad 0 < t < T;$$
$$u(0) = u_0,$$

(1.2)

where the time-dependent operator

$$F(t) : D(F(t)) \subset C[0,1] \longrightarrow C[0,1]$$

is defined by

$$F(t)v = a(x,t)v'' + b(x,t)v' + c(x,t)v + f(x,t);$$

$$v \in D(F(t)) \equiv \{w \in C^2[0,1] : w'(j) = (-1)^j \beta_j(j,t)w(j), j = 0, 1\}.$$

It will be shown by using the theory in Chapter 2 that the equation (1.2) and then the equation (1.1), for $u_0 \in \hat{D}(F(0))$, have a limit solution given by

$$u(t) = \lim_{n \to \infty} \prod_{i=1}^{n} [I - \frac{t}{n} F(i\frac{t}{n})]^{-1} u_0$$

$$= \lim_{\nu \to 0} \prod_{i=1}^{[\frac{t}{\nu}]} [I - \nu F(i\nu)]^{-1} u_0.$$

If $u_0 \in \hat{D}(F(0))$, this $u(t)$ is even a strong solution. It becomes a unique classical solution if more restrictions are imposed on $u_0, a(x,t), b(x,t), c(x,t), f_0(x,t)$, and $\beta_j(x,t)$.

The same results hold true for (1.1) with the Robin boundary condition replaced by the Dirichlet or the Neumann or the periodic one:

- $u(0,t) = 0 = u(1,t)$ (Dirichlet condition).
- $u_x(0,t) = 0 = u_x(1,t)$ (Neumann condition) [15].
- $u(0,t) = u(1,t), \quad u_x(0,t) = u_x(1,t)$ (Periodic condition) [15].

The other case with space dimension greater than one will be the equation, taking the form

$$u_t(x,t) = a_0(x,t)\triangle u(x,t) + \sum_{i=1}^{N} b_i(x)D_i u(x,t)$$

$$+ c(x)u(x,t) + f_0(x,t), \quad (x,t) \in \Omega \times (0,T); \qquad (1.3)$$

$$\frac{\partial}{\partial \hat{n}} u(x,t) + \beta_2(x,t)u(x,t) = 0, \quad x \in \partial\Omega;$$

$$u(x,0) = u_0(x).$$

Here Ω is a bounded, smooth domain in $\mathbb{R}^N$, and $N \geq 2$ is a positive integer; $x = (x_1, x_2, \ldots, x_N)$, and $\triangle u = \sum_{i=1}^{N} \frac{\partial^2}{\partial x_i^2} u$; $u_t = \frac{\partial}{\partial t} u$ and $D_i u = \frac{\partial}{\partial x_i} u$; $\partial\Omega$ is the boundary of Ω, and $\frac{\partial}{\partial \hat{n}} u$ is the outer normal derivative of u.

There are four additional assumptions.

One is that the four functions $a_0(x,t), b_i(x,t), c(x,t)$, and $f_0(x,t)$ are in

$$C^\mu(\overline{\Omega}), \quad 0 < \mu < 1,$$

for all $t \in [0,T]$ where $T > 0$, and are jointly continuous in $x \in \overline{\Omega}$ and $t \in [0,T]$.

The second is that the above four functions satisfy, for $x \in \overline{\Omega}, t, \tau \in [0,T]$, and $\zeta(t)$, a function in t of bounded variation,

$$|a_0(x,t) - a_0(x,\tau)|, \quad |b_i(x,t) - b_i(x,\tau)| \leq |\zeta(t) - \zeta(\tau)|;$$

$$|c(x,t) - c(x,\tau)|, \quad |f_0(x,t) - f_0(x,\tau)| \leq |\zeta(t) - \zeta(\tau)|.$$

The third is that, for all x and t,

$$a_0(x,t), \beta_2(x,t) \geq \delta_0 \quad \text{a positive constant;}$$

$$c(x,t) \leq 0.$$

The fourth is that $\beta_2(x,t)$ is in $C^{1+\mu}(\overline{\Omega})$ for all $t \in [0,T]$ and satisfies, for $x \in \overline{\Omega}, t, \tau \in [0,T]$, and some positive constant M_0,

$$|\beta_2(x,t) - \beta_2(x,\tau)| \le M_0|t - \tau|.$$

The corresponding evolution equation will be

$$\frac{d}{dt}u(t) = G(t)u(t), \quad 0 < t < T; \qquad (1.4)$$
$$u(0) = u_0,$$

in which the time-dependent operator

$$G(t) : D(G(t)) \subset C(\overline{\Omega}) \longrightarrow C(\overline{\Omega})$$

is defined by

$$G(t)v = a_0(x,t)\triangle v + \sum_{i=1}^{N} b_i(x,t)D_iv + c(x,t)v + f_0(x,t);$$

$$v \in D(G(t)) \equiv \{w \in C^{2+\mu}(\overline{\Omega}) : \frac{\partial}{\partial \hat{n}}w + \beta_2(x,t)w = 0 \quad \text{on } \partial\Omega\}.$$

The theory in Chapter 2 will be employed again to prove that, for $u_0 \in \overline{\hat{D}(G(0))}$, the quantity

$$u(t) = \lim_{n\to\infty} \prod_{i=1}^{n}[I - \frac{t}{n}G(i\frac{t}{n})]^{-1}u_0$$

$$= \lim_{\nu\to 0} \prod_{i=1}^{[\frac{t}{\nu}]}[I - \nu G(i\nu)]^{-1}u_0$$

is a limit solution for equation (1.4) or, equivalently, equation (1.3). If $u_0 \in \hat{D}(G(0))$, this $u(t)$ is even a strong solution. Further assumptions on

$$u_0, a_0(x,t), b_i(x,t), c(x,t), f_0(x,t), \text{ and } \beta_2(x,t)$$

will ensure that $u(t)$ is a unique classical solution.

The same results hold true for (1.3) with the Robin boundary condition replaced by the Dirichlet or the Neumann one:

- $u(x,t) = 0$ on $\partial\Omega$ (Dirichlet condition).
- $\frac{\partial}{\partial \hat{n}}u(x,t) = 0$ on $\partial\Omega$ (Neumann condition) [18].

The rest of this chapter is organized as follows. Section 2 states the main results, and Sections 3 and 4 prove the main results.

The material of this chapter is based on [25].

2. Main Results

THEOREM 2.1. *The time-dependent operator $F(t)$ in (1.2) satisfies the four conditions, namely, the dissipativity condition (H1), the range condition (H2), the time-regulating condition (HA), and the embedding condition (HB) in Chapter 2. As a result, it follows from Theorem 2.1, Chapter 2 that, if $u_0 \in \hat{D}(F(0))$, then* *the equation* (1.2) *and then the equation* (1.1) *have a limit solution given by the quantity*

$$u(t) = \lim_{n\to\infty} \prod_{i=1}^{n} [I - \frac{t}{n}F(i\frac{t}{n})]^{-1} u_0$$

$$= \lim_{\nu\to 0} \prod_{i=1}^{[\frac{t}{\nu}]} [I - \nu F(i\nu)]^{-1} u_0.$$

If $u_0 \in \hat{D}(F(0))$, *this* $u(t)$ *is even a strong solution by Theorem 2.2. In that case,* $u(t)$ *also satisfies the middle equation in* (1.1). *It becomes a unique classical solution of* (1.1) *if the two conditions* (S1) *and* (S2) *below are satisfied, and if* u_0 *is such that* $u_0 \in D(F(0))$ *satisfies*

$$F(0)u_0 = [a(x,0)u_0'' + b(x,0)u_0' + c(x,0)u_0 + f_0(x,0)]$$
$$\in D(F(0)).$$

(S1) $a(x,t), b(x,t), c(x,t)$, *and* $f_0(x,t)$ *are the type of the function* $h(x,t)$ *below.* $D_t h(x,t)$ *and* $\triangle h(x,t)$ *exist and are continuous in* x, t, *and* $D_t h(x,t)$ *satisfies, for* $x \in [0,1], t, \tau \in [0,T]$,

$$|D_t h(x,t) - D_t h(x,\tau)| \leq |\zeta(t) - \zeta(\tau)|.$$

Here $D_t h(x,t)$ *is the partial derivative of* $h(x,t)$ *with respect to* t, *and* $\triangle h(x,t)$ *is the second partial derivative of* $h(x,t)$ *with respect to* x.

(S2) $\beta_j(x,t), j = 0,1$ *are twice continuously differentiable in* t, *or weakly satisfy, for* $x \in \{0,1\}, \tau > 0$, *and* $t, t+\tau, t+2\tau \in [0,T]$,

$$|\frac{\beta_j(x,t+2\tau) - 2\beta_j(x,t+\tau) + \beta_j(x,t)}{\tau^2}| \leq M_0,$$

the second difference quotient of $\beta_j(x,t)$ *in* t *being bounded.*

The same results hold true for the equation (1.1) *with the Robin boundary condition replaced by the Dirichlet or the Neumann or the periodic one.*

Remark.

- In order for $F(0)u_0$ to be in $D(F(0))$, more smoothness assumptions should be imposed on the coefficient functions $a(x,0), b(x,0), c(x,0)$, and $f_0(x,0)$.
- The condition $c(x,t) \leq 0$ can be replaced by the condition $c(x,t) \leq \omega$ for some positive constant ω. This is because, in that case, the operator $(F(t) - \omega I)$ is instead considered.

THEOREM 2.2. *The time-dependent operator* $G(t)$ *in* (1.4) *satisfies the four conditions, namely, the dissipativity condition* (H1), *the weaker range condition* (H2)', *the time-regulating condition* (HA), *and the embedding condition* (HB) *in Chapter 2. As a result, it follows from Theorem 2.1, Chapter 2 that, if* $u_0 \in \hat{D}(G(0))$, *then the equation* (1.4) *and then the equation* (1.3) *have a limit solution given by the quantity*

$$u(t) = \lim_{n\to\infty} \prod_{i=1}^{n} [I - \frac{t}{n}G(i\frac{t}{n})]^{-1} u_0$$

$$= \lim_{\nu \to 0} \prod_{i=1}^{[\frac{t}{\nu}]} [I - \nu G(i\nu)]^{-1} u_0.$$

If $u_0 \in \hat{D}(G(0))$, this $u(t)$ is even a strong solution by Theorem 2.2. In that case, $u(t)$ also satisfies the middle equation in (1.3). It becomes a unique classical solution of (1.3) if the two conditions (S3) and (S4) below are satisfied, and if u_0 is such that $u_0 \in D(G(0))$ satisfies

$$G(0)u_0 = [a_0(x,0)\triangle u_0 + \sum_{i=1}^{N} b_i(x,0)D_i u_0 + c(x,0)u_0 + f_0(x,0)]$$

$$\in D(G(0)).$$

(S3) $a_0(x,t), b_i(x,t), c(x,t)$, and $f_0(x,t)$ are the type of the function $h(x,t)$ below. The $(2+\mu)$-th derivative of $h(x,t)$ with respect to x exists and is jointly continuous in x,t. The partial derivative $D_t h(x,t)$ of $h(x,t)$ with respect to t exists and is jointly continuous in x,t, and satisfies, for $x \in \overline{\Omega}, t, \tau \in [0,T]$,

$$|D_t h(x,t) - D_t h(x,\tau)| \leq |\zeta(t) - \zeta(\tau)|.$$

(S4) $\beta_2(x,t)$ is twice continuously differentiable in t, or weakly satisfy, for $x \in \partial\Omega, \tau > 0$, and $t, t+\tau, t+2\tau \in [0,T]$,

$$|\frac{\beta_2(x,t+2\tau) - 2\beta_2(x,t+\tau) + \beta_2(x,t)}{\tau^2}| \leq M_0,$$

the second difference quotient of $\beta_2(x,t)$ in t being bounded.

The same results are valid for the equation (1.3) with the Robin boundary condition replaced by the Dirichlet or the Neumann one.

Remark.

- The condition $G(0)u_0 \in D(G(0))$ requires more smoothness assumptions on the coefficient functions $a_0(x,0), b_i(x,0), c(x,0)$, and $f_0(x,0)$.
- The condition $c(x,t) \leq 0$ can be weakened to $c(x,t) \leq \omega, \omega > 0$, because, in that case, it suffices to consider the operator $(G(t) - \omega I)$.

3. Proof of One Space Dimensional Case

Proof of Theorem 2.1:

PROOF. The proof below assumes the Robin boundary condition, although it is similar with other boundary conditions considered. This proof will be composed of eight steps.

Step 1 It is readily verified as in proving Theorem 2.1 of Chapter 3 and Example 3.1 of Chapter 1 that $F(t)$ satisfies the dissipativity condition (H1).

Step 2. From the theory of ordinary differential equations [5], [24, Corollary 2.13, Chapter 4], the range of $(I - \lambda F(t)), \lambda > 0$, equals $C[0,1]$, so $F(t)$ satisfies the range condition (H2).

Step 3. ($F(t)$ satisfies the time-regulating condition (HA).) Let $g_i(x) \in C[0,1], i = 1,2$, and let

$$v_1 = (I - \lambda F(t))^{-1} g_1;$$

$$v_2 = (I - \lambda F(\tau))^{-1} g_2;$$

where $\lambda > 0$ and $0 \le t, \tau \le T$. Then

$$(v_1 - v_2) - \lambda[a(x,t)(v_1 - v_2)'' + b(x,t)(v_1 - v_2)' + c(x,t)(v_1 - v_2)]$$
$$= \lambda[(a(x,t) - a(x,\tau))v_2'' + (b(x,t) - b(x,\tau))v_2' + (c(x,t) - c(x,\tau))v_2$$
$$+ (f_0(x,t) - f_0(x,\tau))] + (g_1 - g_2);$$
$$\|v_2''\|_\infty \le \delta_0^{-1}[\|F(\tau)v_2\|_\infty + \|b(x,\tau)v_2'\|_\infty + \|c(x,\tau)v_2\|_\infty + \|f_0(x,\tau)\|_\infty];$$
$$(v_1 - v_2)'(0) - \beta_0(0,t)(v_1 - v_2)(0) = [\beta_0(0,t) - \beta_0(0,\tau)]v_2(0);$$
$$(v_1 - v_2)'(1) + \beta_1(1,t)(v_1 - v_2)(1) = -[\beta_1(1,t) - \beta_1(1,\tau)]v_2(1);$$

so there holds, for some function L in the condition (HA),

$$\|v_1 - v_2\|_\infty \le \|g_1 - g_2\|_\infty + \lambda|\zeta(t) - \zeta(\tau)|L(\|v_2\|_\infty)[1 + \|F(\tau)v_2\|_\infty]; \quad \text{or}$$
$$\|v_1 - v_2\|_\infty \le \frac{1}{\delta_0}M_0\|v_2\|_\infty|t - \tau|,$$

thus proving the condition (HA), where

$$F(\tau)v_2 = \frac{v_2 - g_2}{\lambda} = \frac{[I - \lambda F(\tau)]^{-1}g_2 - g_2}{\lambda}.$$

This is because of the interpolation inequality [1], [13, page 135]

$$\|v_2'\|_\infty \le \epsilon\|v_2''\|_\infty + C(\epsilon)\|v_2\|_\infty \tag{3.1}$$

for each $\epsilon > 0$ and for some constant $C(\epsilon)$, and because, as in solving Example 3.1, Chapter 1, the maximum principle applies; that is, there is an $x_0 \in [0,1]$ such that

$$\|v_1 - v_2\|_\infty = |(v_1 - v_2)(x_0)|,$$

that, for $x_0 \in (0,1)$,

$$(v_1 - v_2)'(x_0) = 0;$$
$$(v_1 - v_2)(x_0)(v_1 - v_2)''(x_0) \le 0,$$

and that, for $x_0 \in \{0,1\}$,

$$(v_1 - v_2)'(0) \le 0 \quad \text{or} \ge 0, \text{ according as } (v_1 - v_2)(0) > 0 \text{ or} < 0;$$
$$(v_1 - v_2)'(1) \ge 0 \quad \text{or} \le 0, \text{ according as } (v_1 - v_2)(1) > 0 \text{ or} < 0.$$

Step 4. $(F(t)$ satisfies the embedding condition (HB).) [20] Let $t_n \in [0,T]$ converge to t, $v_n \in D(F(t_n))$ converge to v in $C[0,1]$, and $\|F(t_n)v_n\|_\infty$ be uniformly bounded. It will be shown that, for each η in the self-dual space $L^2(0,1) = (L^2(0,1))^*$, $\eta(F(t)v)$ exists and

$$|\eta(F(t_n)v_n) - \eta(F(t)v)| \longrightarrow 0.$$

Here $(C[0,1]; \|\cdot\|_\infty)$ is continuously embedded into $L^2(0,1); \|\cdot\|)$.

Since $\|v_n\|_\infty$ and $\|F(t_n)v_n\|_\infty$ are uniformly bounded, so is $\|v_n\|_{C^2[0,1]}$ by the interpolation inequality [1], [13, page 135]. Hence, by Ascoli-Arzela theorem [33], a subsequence of v_n and then itself converge in $C^1[0,1]$ to v. Also, v_n is uniformly bounded in the Hilbert space $W^{2,2}(0,1)$, whence, by the Alaoglu theorem [36], a subsequence of v_n and then itself converge weakly to v [36]. It follows that, for each $\eta \in L^2(0,1)$,

$$|\eta(F(t_n)v_n) - \eta(F(t)v)| \longrightarrow 0,$$

because

$$\int_0^1 a(x,t)(v'' - v_n'')\eta \, dx + \int_0^1 [a(x,t) - a(x,t_n)]v_n''\eta \, dx$$

$$+ \int_0^1 b(x,t)(v' - v_n')\eta \, dx + \int_0^1 [b(x,t) - b(x,t_n)]v_n'\eta \, dx$$

$$+ \int_0^1 c(x,t)(v - v_n)\eta \, dx + \int_0^1 [c(x,t) - c(x,t_n)]v_n\eta \, dx$$

$$+ \int_0^1 [f_0(x,t) - f_0(x,t_n)]\eta \, dx$$

$$\longrightarrow 0.$$

Therefore $F(t)$ satisfies the embedding condition (HB).

Step 5. ($u(t)$ for $u_0 \in \hat{D}(F(0))$ satisfies the middle equation in (1.1).) Consider the discretized equation

$$u_i - \nu F(t_i)u_i = u_{i-1},$$
$$u_i \in D(F(t_i)), \tag{3.2}$$

where $u_0 \in \hat{D}(F(0)), i = 1, 2, \ldots, n$, $n \in \mathbb{N}$ is large, and $\nu > 0$ is such that

$$\nu\omega < 1 \text{ and } 0 \le t_i = i\nu \le T.$$

Here

$$u_i = \prod_{k=1}^i [I - \nu F(t_k)]^{-1} u_0$$

exists uniquely by the range condition (H2) and the dissipativity condition (H1). For convenience, we also define

$$u_{-1} = u_0 - \nu F(0)u_0.$$

Now, for each $t \in [0,T)$, we have $t \in [t_i, t_{i+1})$ for some i, so $i = [\frac{t}{\nu}]$. It follows from Theorem 2.1 in Chapter 2 that, for each above t with the corresponding i,

$$\lim_{\nu \to 0} u_i = \lim_{\nu \to 0} \prod_{k=1}^{[\frac{t}{\nu}]} [I - \nu F(t_k)]^{-1} u_0$$

$$= \lim_{n \to \infty} \prod_{k=1}^n [I - \frac{t}{n}F(k\frac{t}{n})]^{-1} u_0$$

$$\equiv u(t)$$

exists.

On the other hand, by utilizing Proposition 4.2 in Section 4 of Chapter 2, we have

$$\|u_i\|_\infty;$$
$$\|a(x,t_i)u_i'' + b(x,t_i)u_i' + c(x,t_i)u_i + f_0(x,t_i)\|_\infty$$
$$= \|F(t_i)u_i\|_\infty = \|\frac{u_i - u_{i-1}}{\nu}\|_\infty;$$

are uniformly bounded. Those, in turn, result in a bound for $\|u_i\|_{C^2[0,1]}$ by the interpolation inequality [1], [13, page 135]. Therefore it follows from Ascoli-Arzela theorem [33] that a subsequence of u_i and then itself converge in $C^1[0,1]$ to a limit, as $\nu \longrightarrow 0$. This limit equals $u(t)$ as shown above. Consequently, $u(t)$ satisfies the middle equation in (1.1), as u_i does so.

Step 6. (Further estimates of u_i under the conditions (S1), (S2), and $u_0 \in D(F(0))$ with $F(0)u_0 \in D(F(0))$) Because of $F(0)u_0 \in D(F(0))$, the u_i in Step 5 satisfies

$$u_i - \nu[a(x,t_i)u_i'' + b(x,t_i)u_i' + c(x,t_i)u_i + f_0(x,t_i)] = u_{i-1},$$
$$i = 0, 1, \ldots;$$
$$u_i'(0) = \beta_0(0,t_i)u_i(0), \quad u_i'(1) = -\beta_1(1,t_i)u_i(1),$$
$$i = -1, 0, 1, \ldots.$$

From this, it follows, on letting $v_i = \frac{u_i - u_{i-1}}{\nu}$ for $i = 0, 1, \ldots$, that

$$v_i - \nu[a(x,t_i)v_i'' + b(x,t_i)v_i' + c(x,t_i)v_i + g(x,\nu,t_i)] = v_{i-1},$$
$$i = 1, 2, \ldots;$$
$$v_i'(0) - \beta_0(0,t_i)v_i(0) = \frac{\beta_0(0,t_i) - \beta_0(0,t_{i-1})}{\nu}u_{i-1}(0);$$
$$v_i'(1) + \beta_1(1,t_i)v_i(1) = -\frac{\beta_1(1,t_i) - \beta_1(1,t_{i-1})}{\nu}u_{i-1}(1),$$

where, with $t_{i-1} = t_i - \nu$,

$$g(x,\nu,t_i) = g(x,\nu,t_i,t_{i-1})$$
$$= \frac{a(x,t_i) - a(x,t_{i-1})}{\nu}u_{i-1}'' + \frac{b(x,t_i) - b(x,t_{i-1})}{\nu}u_{i-1}'$$
$$+ \frac{c(x,t_i) - c(x,t_{i-1})}{\nu}u_{i-1} + \frac{f_0(x,t_i) - f_0(x,t_{i-1})}{\nu}.$$

Here, for convenience, we also define

$$v_{-1} = v_0 - \nu[a(x,t_0)v_0'' + b(x,t_0)v_0' + c(x,t_0)v_0 + g(x,\nu,t_0)];$$
$$t_{-1} = 0;$$

for which $g(x,\nu,t_0) = g(x,\nu,0) = 0$.

Thus, either from Corollary 4.3 or from the proof of Proposition 4.1 and the results in and the proof of Proposition 4.2 in Section 4 of Chapter 2, we have

$$\|a(x,t_i)v_i'' + b(x,t_i)v_i' + c(x,t_i)v_i + g(x,\nu,t_i)\|_\infty$$
$$= \|\frac{v_i - v_{i-1}}{\nu}\|_\infty, \quad i = 0, 1, \ldots;$$

is uniformly bounded, whence so are

$$\|v_i\|_{C^2[0,1]} = \|\frac{u_i - u_{i-1}}{\nu}\|_{C^2[0,1]}$$
$$= \|a(x,t_i)u_i'' + b(x,t_i)u_i' + c(x,t_i)u_i + f_0(x,t_i)\|_{C^2[0,1]},$$
$$i = 0, 1, \ldots;$$
$$\|u_i\|_{C^4[0,1]}, \quad i = 0, 1, \ldots,$$

as in Step 5. This is because those v_i's above, $i = -1, 0, 1, \ldots$, satisfy the conditions (C1), (C2), and (C3) in Corollary 4.3, that is, the conditions ((4.3) or (4.4)), ((4.5) or (4.6)), and ((4.7) or (4.8)) in Section 4 of Chapter 2. A proof of it follows from applying (3.1), the maximum principle argument in Step 3, and the fact that the quantity $\frac{\|u_i - u_{i-1}\|_\infty}{\nu}$ in Step 5 is bounded. Here it is to be observed that $\|g(x, \nu, t_i)\|_\infty$ is uniformly bounded, and that

$$\frac{u''_{i-1} - u''_{i-2}}{\nu} = v''_{i-1}$$

$$= \frac{1}{a(x, t_{i-1})} \left[\frac{v_{i-1} - v_{i-2}}{\nu} - b(x, t_{i-1})v'_{i-1} \right.$$

$$\left. - c(x, t_{i-1})v_{i-1} - g(x, \nu, t_{i-1}) \right].$$

Step 7. (Existence of a solution) Now that, from Step 6, $\|u_i\|_{C^4[0,1]}, i = 2, 3, \ldots$, is uniformly bounded, it follows from the Ascoli-Arzela theorem [**33**], as in Step 5, that a subsequence of u_i and then itself, through the discretized equation (3.2), converge in $C^3[0, 1]$ to the limit $u(t)$, as $\nu \longrightarrow 0$. Therefore $u(t)$ is a classical solution.

Step 8. (Uniqueness of a solution) This proceeds as in Step 5 in the proof of Example 3.2, Chapter 1.

The proof is complete.

$\square$

4. Proof of Higher Space Dimensional Case

Proof of Theorem 2.2:

PROOF. The proof below is composed of eight steps. The proof is proceeded with the Robin boundary condition assumed, while it is similar if other boundary conditions are considered.

Step 1 It is readily verified as in proving Example 3.2 of Chapter 1 and Theorem 2.2 of Chapter 3 that $G(t)$ satisfies the dissipativity condition (H1).

Step 2. From the theory of linear, elliptic partial differential equations [**13**], the range of $(I - \lambda G(t)), \lambda > 0$, equals $C^\mu(\overline{\Omega})$, so $G(t)$ satisfies the weaker range condition $(H2)'$ because of $C^\mu(\overline{\Omega}) \supset D(G(t))$ for all t.

Step 3. $(G(t)$ satisfies the time-regulating condition (HA).) Let $g_i(x) \in C^\mu(\overline{\Omega}), i = 1, 2$, and let

$$v_1 = (I - \lambda G(t))^{-1} g_1;$$

$$v_2 = (I - \lambda G(\tau))^{-1} g_2;$$

where $\lambda > 0$ and $0 \le t, \tau \le T$. Then

$$(v_1 - v_2) - \lambda \left[a_0(x, t)\triangle(v_1 - v_2) + \sum_{i=1}^N b_i(x, t)D_i(v_1 - v_2) + c(x, t)(v_1 - v_2) \right]$$

$$= \lambda \left[(a_0(x, t) - a_0(x, \tau))\triangle v_2 + \left(\sum_{i=1}^n (b_i(x, t) - b_i(x, \tau))D_i v_2 \right. \right.$$

$$\left. \left. + (c(x, t) - c(x, \tau))v_2 + (f_0(x, t) - f_0(x, \tau)) \right] + (g_1 - g_2), \quad x \in \Omega;$$

$$\|\triangle v_2\|_\infty \le \delta^{-1} \left[\|G(\tau)v_2\|_\infty + \sum_{i=1}^N \|b_i(x, \tau)D_i v_2\|_\infty \right.$$

$$\left. + \|c(x, \tau)v_2\|_\infty + \|f_0(x, \tau)\|_\infty;$$

$$\frac{\partial(v_1 - v_2)}{\partial \hat{n}} + \beta_2(x, t)(v_1 - v_2) = -(\beta_2(x, t) - \beta_2(x, \tau))v_2, \quad x \in \partial\Omega;$$

so there holds, for some function L in the condition (HA),

$$\|v_1 - v_2\|_\infty \le \|g_1 - g_2\|_\infty + \lambda|\zeta(t) - \zeta(\tau)|L(\|v_2\|_\infty)[1 + \|G(\tau)v_2\|_\infty]; \quad \text{or}$$

$$\|v_1 - v_2\|_\infty \le \frac{1}{\delta}M_0\|v_2\|_\infty|t - \tau| \le L(\|v_2\|_\infty)|t - \tau|,$$

where

$$G(\tau)v_2 = \frac{v_2 - g_2}{\lambda} = \frac{[I - \lambda G(\tau)]^{-1}g_2 - g_2}{\lambda}.$$

This is because, as in proving the dissipativity condition (H1) in Step 1, the maximum principle argument applies, that is, there is an $x_0 \in \overline{\Omega}$ such that

$$\|v_1 - v_2\|_\infty = |(v_1 - v_2)(x_0)|,$$

that, for $x_0 \in \Omega$,

$$D(v_1 - v_2)(x_0) = 0;$$
$$(v_1 - v_2)(x_0)\Delta(v_1 - v_2)(x_0) \le 0,$$

and that, for $x_0 \in \partial\Omega$,

$$\frac{\partial(v_1 - v_2)}{\partial\hat{n}}(x_0) \ge 0 \quad \text{or} \le 0 \text{ according as } (v_1 - v_2)(x_0) > 0 \text{ or } < 0.$$

Here, to derive, for some constants c_1 and c_2,

$$\|Dv_2\|_\infty \le \|v_2\|_{C^{1+\mu}(\overline{\Omega})}$$
$$\le c_1\|\frac{v_2 - g_2}{\lambda}\|_\infty + c_2\|v_2\|_\infty, \tag{4.1}$$

we used the integral representation of v_2, with the Green's function $Z(x, y)$ of the second kind [29],

$$v_2 = -\int_{\partial\Omega} Z(x, y)[-\beta_2(y, \tau)v_2]\,d\sigma_y$$

$$+ \int_\Omega Z(x, y)a_0(y, \tau)^{-1}[\frac{v_2 - g_2}{\lambda} - \sum_{i=1}^{N} b_i(y, \tau)D_iv_2 - c(y, \tau)v_2]\,dy,$$

differentiated the above v_2 for $(1 + \mu)$ times, with respect to the variable x, to obtain, for some constants d_1 and d_2,

$$\|Dv_2\|_{C^\mu(\overline{\Omega})} \le d_1\|\frac{v_2 - g_2}{\lambda}\|_\infty + d_2(\|Dv_2\|_\infty + \|v_2\|_\infty),$$

and then employed the interpolation inequality [1], [13, page 135]

$$\|Dv_2\|_\infty \le \epsilon\|Dv_2\|_{C^\mu(\overline{\Omega})} + C(\epsilon)\|v_2\|_\infty$$

for each $\epsilon > 0$ and for some constant $C(\epsilon)$. Here to be noticed were, for some constants b_1 and b_2,

$$|D_iZ(x, y)| \le b_1|x - y|^{1-N};$$
$$|D_{ij}Z(x, y)| \le b_2|x - y|^{-N};$$
$$\int_\Omega |x - y|^{N(\mu-1)}\,dx \quad \text{is finite for } 0 < \mu < 1 \text{ [13, page 159]}.$$

Step 4. ($G(t)$ satisfies the embedding condition (HB).) [20] Let $t_n \in [0, T]$ converge to t, $v_n \in D(G(t_n))$ converge to v in $C(\overline{\Omega})$, and $\|G(t_n)v_n\|_\infty$ be uniformly bounded. It will be shown that, for each η in the self-dual space $L^2(\Omega) = (L^2(\Omega))^*$, $\eta(G(t)v)$ exists and

$$|\eta(G(t_n)v_n) - \eta(G(t)v)| \longrightarrow 0.$$

Here $(C(\overline{\Omega}); \|\cdot\|_\infty)$ is continuously embedded into $L^2(\Omega); \|\cdot\|)$.

Since $\|v_n\|_\infty$ and $\|G(t_n)v_n\|_\infty$ are uniformly bounded, so is $\|v_n\|_{C^{1+\lambda}(\overline{\Omega})}$ for any $0 < \lambda < 1$, using the proof of (4.1). (Alternatively, uniformly bounded are $\|v_n\|_{W^{2,p}(\Omega)}$ for any $p \geq 2$ and then $\|v_n\|_{C^{1+\lambda}(\overline{\Omega})}$ for any $0 < \lambda < 1$, on using the L^p elliptic estimates [37] and the Sobolev embedding theorem [1, 13].) Hence, by the Ascoli-Arzela theorem [33], a subsequence of v_n and then itself converge in $C^1(\overline{\Omega})$ to v. Also, v_n is uniformly bounded in the Hilbert space $W^{2,2}(\Omega)$, whence, by the Alaoglu theorem [36], a subsequence of v_n and then itself converge weakly to v [36]. It follows that, for each $\eta \in L^2(\Omega)$,

$$|\eta(G(t_n)v_n) - \eta(G(t)v)| \longrightarrow 0,$$

because

$$\int_\Omega a_0(x,t)(\triangle v - \triangle v_n)\eta \, dx + \int_\Omega [a_0(x,t) - a_0(x,t_n)]\triangle v_n \eta \, dx$$

$$+ \int_\Omega \sum_{i=1}^N b_i(x,t)(D_i v - D_i v_n)\eta \, dx + \int_\Omega \sum_{i=1}^N [b_i(x,t) - b_i(x,t_n)]D_i v_n \eta \, dx$$

$$+ \int_\Omega c(x,t)(v - v_n)\eta \, dx + \int_\Omega [c(x,t) - c(x,t_n)]v_n \eta \, dx$$

$$+ \int_\Omega [(f_0(x,t_n) - f_0(x,t))]\eta \, dx|$$

$$\longrightarrow 0.$$

Therefore $G(t)$ satisfies the embedding condition (HB).

Step 5. ($u(t)$ for $u_0 \in \hat{D}(G(0))$ satisfies the middle equation in (1.3).) Consider the discretized equation

$$u_i - \nu G(t_i)u_i = u_{i-1},$$
$$u_i \in D(G(t_i)), \tag{4.2}$$

where $u_0 \in \hat{D}(G(0))$, $i = 1, 2, \ldots, n$, $n \in \mathbb{N}$ is large, and $\nu > 0$ is such that

$$\nu\omega < 1 \text{ and } 0 \leq t_i = i\nu \leq T.$$

Here

$$u_i = \prod_{k=1}^i [I - \nu G(t_k)]^{-1} u_0$$

exists uniquely by the range condition (H2) and the dissipativity condition (H1). For convenience, we also define

$$u_{-1} = u_0 - \nu G(0)u_0.$$

Now, for each $t \in [0, T)$, we have $t \in [t_i, t_{i+1})$ for some i, so $i = [\frac{t}{\nu}]$. It follows from Theorem 2.1 in Chapter 2 that, for each above t with the corresponding i,

$$\lim_{\nu \to 0} u_i = \lim_{\nu \to 0} \prod_{k=1}^{[\frac{t}{\nu}]} [I - \nu G(t_k)]^{-1} u_0$$

$$= \lim_{n \to \infty} \prod_{k=1}^{n} [I - \frac{t}{n} G(k\frac{t}{n})]^{-1} u_0$$

$$\equiv u(t)$$

exists.

On the other hand, by utilizing Proposition 4.2 in Section 4 of Chapter 2, we have

$$\|u_i\|_\infty;$$

$$\|a_0(x, t_i) \triangle u_i + \sum_{j=1}^{N} b_j(x, t_i) D_j u_i + c(x, t_i) u_i + f_0(x, t_i)\|_\infty$$

$$= \|G(t_i) u_i\|_\infty = \|\frac{u_i - u_{i-1}}{\nu}\|_\infty;$$

are uniformly bounded, whence so is $\|u_i\|_{C^{1+\lambda}(\overline{\Omega})}$ for any $0 < \lambda < 1$, using the proof of (4.1). (Alternatively, those, in turn, result in a bound for $\|u_i\|_{W^{2,p}(\Omega)}$ for any $p \geq 2$, by the L^p elliptic estimates [37]. Hence, a bound exists for $\|u_i\|_{C^{1+\eta}(\overline{\Omega})}$ for any $0 < \eta < 1$, as a result of the Sobolev embedding theorem [1, 13].) Therefore it follows from Ascoli-Arzela theorem [33] that a subsequence of u_i and then itself converge in $C^{1+\mu}(\overline{\Omega})$ to a limit, as $\nu \longrightarrow 0$. This limit equals $u(t)$ as shown above. Consequently, $u(t)$ satisfies the middle equation in (1.3), as u_i does so.

Step 6. (Further estimates of u_i under the conditions (S3), (S4), and $u_0 \in D(G(0))$ with $G(0)u_0 \in D(G(0))$) Because of $G(0)u_0 \in D(G(0))$, the u_i in Step 5 satisfies

$$u_i - \nu[a_0(x, t_i) \triangle u_i + \sum_{j=1}^{N} b_j(x, t_i) D_j u_i$$

$$+ c(x, t_i) u_i + f_0(x, t_i)] = u_{i-1}, \quad x \in \Omega, \quad i = 0, 1, \dots;$$

$$\frac{\partial}{\partial \hat{n}} u_i(x) + \beta_2(x, t_i) u_i(x) = 0, \quad x \in \partial\Omega, \quad i = -1, 0, 1, \dots.$$

From this, it follows, on letting $w_i = \frac{u_i - u_{i-1}}{\nu}$ for $i = 0, 1, \dots$, that

$$w_i - \nu[a_0(x, t_i) \triangle w_i + \sum_{j=1}^{N} b_j(x, t_i) D_j w_i$$

$$+ c(x, t_i) w_i + g(x, \nu, t_i)] = w_{i-1}, \quad x \in \Omega, \quad i = 1, 2, \dots;$$

$$\frac{\partial w_i}{\partial \hat{n}} + \beta_2(x, t_i) w_i = -\frac{\beta_2(x, t_i) - \beta_2(x, t_{i-1})}{\nu} u_{i-1},$$

$$x \in \partial\Omega, \quad i = 0, 1, \dots;$$

where, with $t_{i-1} = t_i - \nu$,

$$g(x, \nu, t_i) = g(x, \nu, t_i, t_{i-1})$$

$$= \frac{a_0(x, t_i) - a_0(x, t_{i-1})}{\nu} \triangle u_{i-1} + \sum_{j=1}^{N} \frac{b_j(x, t_i) - b_j(x, t_{i-1})}{\nu} D_j u_{i-1}$$

$$+ \frac{c(x, t_i) - c(x, t_{i-1})}{\nu} u_{i-1} + \frac{f_0(x, t_i) - f_0(x, t_{i-1})}{\nu}, \quad i = 0, 1, \dots.$$

Here, for convenience, we also define

$$w_{-1} = w_0 - \nu[a_0(x,t_0)\triangle w_0 + \sum_{j=1}^{N} b_j(x,t_0)D_j w_0$$
$$+ c(x,t_0)w_0 + g(x,\nu,t_0))];$$
$$t_{-1} = 0;$$

for which $g(x,\nu,t_0) = g(x,\nu,0) = 0$.

Hence, either from Corollary 4.3 or from the proof of Proposition 4.1 and from both the results in Proposition 4.2 and the proof of Proposition 4.2 in Section 4 of Chapter 2, we have

$$\|a_0(x,t_i)\triangle w_i + \sum_{j=1}^{N} b_j(x,t_i)D_j w_i + c(x,t_i)w_i + g(x,\nu,t_i)\|_\infty$$

$$= \|\frac{w_i - w_{i-1}}{\nu}\|_\infty, \quad i=0,1,\ldots;$$
$$\|w_i\|_{C^{1+\eta}(\overline{\Omega})}, \quad 0<\eta<1, \quad i=0,1,\ldots;$$

are uniformly bounded, as in Step 5, where

$$w_i = \frac{u_i - u_{i-1}}{\nu}$$

$$= a_0(x,t_i)\triangle u_i + \sum_{j=1}^{N} b_j(x,t_i)D_j u_i + c(x,t_i)u_i + f_0(x,t_i);$$

hence, so is

$$\|u_i\|_{C^{3+\eta}(\overline{\Omega})}, \quad i=0,1,\ldots$$

by the Schauder global regularity theorem [**13**, page 111]. This is because those w_i's, $i=-1,0,1,\ldots$, satisfy the conditions (C1), (C2), and (C3) in Corollary 4.3, that is, the conditions ((4.3) or (4.4)), ((4.5) or (4.6)), and ((4.7) or (4.8)) in Section 4 of Chapter 2, for which employed were (4.1) and both the maximum principle argument in Step 3 and the boundedness of $\frac{\|u_i-u_{i-1}\|_\infty}{\nu}$ in Step 5. Here it is to be observed that $\|g(x,\nu,t_i)\|_\infty$ is uniformly bounded, and that

$$\frac{\triangle u_{i-1} - \triangle u_{i-2}}{\nu} = \triangle w_{i-1}$$

$$= \frac{1}{a_0(x,t_{i-1})}[\frac{w_{i-1} - w_{i-2}}{\nu} - \sum_{j=1}^{N} b_j(x,t_{i-1})D_j w_{j-1}$$

$$- c(x,t_{i-1})w_{i-1} - g(x,\nu,t_{i-1})].$$

Step 7. (Existence of a solution) Now that, from Step 6, $\|u_i\|_{C^{3+\eta}(\overline{\Omega})}, i = 2,3,\ldots$, is uniformly bounded, it follows from the Ascoli-Arzela theorem [**33**], as in Step 5, that a subsequence of u_i and then itself, through the discretized equation (3.4), converge in $C^{3+\mu}(\overline{\Omega})$ to the limit $u(t)$, as $\nu \longrightarrow 0$. Therefore $u(t)$ is a classical solution.

Step 8. (Uniqueness of a solution) This proceeds as in Step 5 in the proof of Example 3.2, Chapter 1.

The proof is complete.

$\square$

CHAPTER 6

Nonlinear Non-autonomous Parabolic Equations (I)

1. Introduction

In this chapter, nonlinear non-autonomous, parabolic initial-boundary value problems will be solved with the aid of the results in Chapter 2. The obtained solutions will be strong ones under suitable assumptions.

Thus, by extending the nonlinear autonomous cases in Chapter 4 and the linear non-autonomous cases in Chapter 5, we shall consider the nonlinear, non-autonomous equation with the nonlinear Robin boundary condition, as well as its higher space dimensional analogue

$$u_t(x,t) = \alpha(x,t,u_x)u_{xx}(x,t) + g(x,t,u,u_x),$$
$$(x,t) \in (0,1) \times (0,T);$$
$$u_x(0,t) \in \beta_0(u(0,t)), \quad u_x(1,t) \in -\beta_1(u(1,t)); \tag{1.1}$$
$$u(x,0) = u_0(x).$$

Here the assumptions are made:

- $\beta_0, \beta_1 : \mathbb{R} \longrightarrow \mathbb{R}$, are multi-valued, maximal monotone functions with $0 \in \beta_0(0) \cap \beta_1(0)$.
- $\alpha(x,t,p)$ and $g(x,t,z,p)$ are real-valued, continuous functions of their arguments $x \in [0,1], t \in [0,T], z \in \mathbb{R}$, and $p \in \mathbb{R}$. Here $T > 0$.
- $\alpha(x,t,p)$ is greater than or equal to some positive constant δ_0 for all its arguments x, t, and p.
- $g(x,t,z,p)$ is monotone non-increasing in z for each x,t, and p; that is,

$$(z_2 - z_1)[g(x,t,z_2,p) - g(x,t,z_1,p)] \leq 0.$$

- $g(x,t,z,p)$ is of at most linear growth in p, that is, for some positive, continuous function $M_0(x,t,z)$,

$$|g(x,t,z,p)| \leq M_0(x,t,z)(1 + |p|).$$

- The following are true for some continuous, positive functions N_0 and N_{01} and for some continuous function ζ of bounded variation:

$$|\alpha(x,t,p) - \alpha(x,\tau,p)|/\alpha(x,\tau,p) \leq |\zeta(t) - \zeta(\tau)|N_0(x,t,\tau);$$
$$|g(x,t,z,p) - g(x,\tau,z,p)| \leq |\zeta(t) - \zeta(\tau)|N_{01}(x,t,\tau,z)(1 + |p|).$$

Here it is to be noted that assuming linear growth of $g(x,t,z,p)$ in p is stronger than assuming that of $g(x,z,p)/\alpha(x,p)$ in p in Chapter 4.

(1.1) will be written as the nonlinear evoution

$$\frac{d}{dt}u(t) = H(t)u(t), \quad t \in (0,T);$$

$$u(0) = u_0,$$

(1.2)

where the nonlinear, multi-valued operator

$$H(t) : D(H(t)) \subset C[0,1] \longrightarrow C[0,1]$$

is defined by

$$H(t)v = \alpha(x,t,v')v'' + g(x,t,v,v');$$

$$v \in D(H(t)) \equiv \{w \in C^2[0,1] : w'(j) \in (-1)^j \beta_j w(j), j = 0,1\}.$$

It will be shown by using the theory in Chapter 2 that the equation (1.2) and then the equation (1.1), for $u_0 \in \overline{\hat{D}(H(0))}$, have a limit solution given by

$$u(t) = \lim_{n \to \infty} \prod_{i=1}^{n} I - \frac{t}{n} H(i\frac{t}{n})]^{-1} u_0$$

$$= \lim_{\nu \to 0} \prod_{i=1}^{[\frac{t}{\nu}]} I - \nu H(i\nu)]^{-1} u_0.$$

If $u_0 \in \hat{D}(H(0))$, this $u(t)$ is even a strong solution.

The higher space dimensional analogue of (1.1) will be of the form

$$u_t(x,t) = \alpha_0(x,t,Du)\Delta u(x,t) + g_0(x,t,u,Du),$$

$$(x,t) \in \Omega \times (0,T);$$

$$\frac{\partial}{\partial \hat{n}} u(x,t) + \beta(x,t,u) = 0, \quad x \in \partial\Omega;$$

$$u(x,0) = u_0(x);$$

(1.3)

in which seven assumptions are made.

- Ω is a bounded smooth domain in $\mathbb{R}^N, N \geq 2$, and $\partial\Omega$ is the boundary of Ω.
- $\hat{n}(x)$ is the unit outer normal to $x \in \partial\Omega$, and μ is a real number such that $0 < \mu < 1$.
- $\alpha_0(x,t,p) \in C^{1+\mu}(\overline{\Omega} \times \mathbb{R}^N)$ is true for each $t \in [0,T]$ where $T > 0$, and is continuous in all its arguments. Furthermore, $\alpha_0(x,t,p) \geq \delta_1 > 0$ is true for all x,p, and all $t \in [0,T]$, and for some constant $\delta_1 > 0$.
- $g_0(x,t,z,p) \in C^{1+\mu}(\overline{\Omega} \times \mathbb{R} \times \mathbb{R}^N)$ is true for each $t \in [0,T]$, is continuous in all its arguments, and is monotone non-increasing in z for each t,x, and p.
- $g_0(x,t,z,p)$ is of at most linear growth in p, that is,

$$|g_0(x,t,z,p)| \leq M_1(x,t,z)(1+|p|)$$

for some positive, continuous function M_1 and for all $t \in [0,T]$.

- $\beta(x,t,z) \in C^{2+\mu}(\Omega \times \mathbb{R})$ is true for each $t \in [0,T]$, is continuous in all its arguments, and is strictly monotone increasing in z so that $\beta_z \geq \delta_1 > 0$ for the constant $\delta_1 > 0$.

- The following are true for some continuous, positive functions N_1, N_2, N_3 and for some continuous function ζ of bounded variation:

$$|\alpha_0(x,t,p) - \alpha_0(x,\tau,p)|/\alpha_0(x,\tau,p) \le |\zeta(t) - \zeta(\tau)|N_1(x);$$
$$|g_0(x,t,z,p) - g_0(x,\tau,z,p)| \le |\zeta(t) - \zeta(\tau)|N_2(x,z)(1+|p|);$$
$$|\beta(x,t,z) - \beta(x,\tau,z)| \le |t - \tau|N_3(x,z).$$

Here again, it is to be noted that assuming linear growth of $g_0(x,t,z,p)$ in p is stronger than assuming that of $g_0(x,z,p)/\alpha_0(x,p)$ in p in Chapter 4.

The corresponding nonlinear evolution equation will be

$$\frac{d}{dt}u(t) = J(t)u(t), \quad t \in (0,T);$$
$$u(0) = u_0, \tag{1.4}$$

in which the time-dependent, nonlinear operator

$$J(t) : D(J(t)) \subset C(\overline{\Omega}) \longrightarrow C(\overline{\Omega})$$

is defined by

$$J(t)v = \alpha_0(x,t,Dv)\triangle v + g_0(x,t,v,Dv);$$

$$v \in D(J(t)) \equiv \{w \in C^{2+\mu}(\overline{\Omega}) : \frac{\partial}{\partial \hat{n}}w + \beta(x,t,w) = 0 \quad \text{on } \partial\Omega\}.$$

The theory in Chapter 2 will be employed again to prove that, for $u_0 \in \overline{D(J(0))}$, the quantity

$$u(t) = \lim_{n\to\infty} \prod_{i=1}^{n}[I - \frac{t}{n}J(i\frac{t}{n})]^{-1}u_0$$

$$= \lim_{\nu\to 0} \prod_{i=1}^{[\frac{t}{\nu}]}[I - \nu J(i\nu)]^{-1}u_0$$

is a limit solution for equation (1.4) or, equivalently, equation (1.3). If $u_0 \in \hat{D}(J(0))$, this $u(t)$ is even a strong solution.

The rest of this chapter is organized as follows. Section 2 states the main results, and Sections 3 and 4 prove the main results.

The material of this chapter is based on [25].

2. Main Results

THEOREM 2.1. *The time-dependent operator $H(t)$ in (1.2) satisfies the four conditions, namely, the dissipativity condition (H1), the range condition (H2), the time-regulating condition (HA), and the embedding condition (HB) in Chapter 2. As a result, it follows from Theorem 2.1, Chapter 2 that, if $u_0 \in \hat{D}(H(0))$, then the equation (1.2) and then the equation (1.1) have a limit solution given by the quantity*

$$u(t) = \lim_{n\to\infty} \prod_{i=1}^{n}[I - \frac{t}{n}H(i\frac{t}{n})]^{-1}u_0$$

$$= \lim_{\nu\to 0} \prod_{i=1}^{[\frac{t}{\nu}]}[I - \nu H(i\nu)]^{-1}u_0.$$

If $u_0 \in \hat{D}(H(0))$, this $u(t)$ is even a strong solution by Theorem 2.2. In that case, $u(t)$ also satisfies the middle equation in (1.1).

Remark.

- The condition

$$(z_2 - z_1)[g(x, z_2, p) - g(x, z_1, p)] \leq 0$$

can be replaced by the condition

$$(z_2 - z_1)[g(x, z_2, p) - g(x, z_1, p)] \leq \omega(z_2 - z_1)^2$$

for some positive constant ω. This is because, in that case, the operator $(H - \omega I)$ is instead considered.

THEOREM 2.2. *The time-dependent operator $J(t)$ in (1.4) satisfies the four conditions, namely, the dissipativity condition (H1), the weaker range condition (H2)′, the time-regulating condition (HA), and the embedding condition (HB) in Chapter 2. As a result, it follows from Theorem 2.1, Chapter 2 that, if $u_0 \in \hat{D}(J(0))$, then equation (1.4) or, equivalently, equation (1.3) has a limit solution given by the quantity*

$$u(t) = \lim_{n \to \infty} \prod_{i=1}^{n} [I - \frac{t}{n} J(i\frac{t}{n})]^{-1} u_0$$

$$= \lim_{\nu \to 0} \prod_{i=1}^{[\frac{t}{\nu}]} [I - \nu J(i\nu)]^{-1} u_0.$$

If $u_0 \in \hat{D}(J(0))$, this $u(t)$ is even a strong solution by Theorem 2.2. In that case, $u(t)$ also satisfies the middle equation in (1.3).

Remark.

- The condition

$$(z_2 - z_1)[g_0(x, z_2, p) - g_0(x, z_1, p)] \leq 0$$

can be weakened to

$$(z_2 - z_1)[g_0(x, z_2, p) - g_0(x, z_1, p)] \leq \omega(z_2 - z_1)^2$$

for some $\omega > 0$, because, in that case, it suffices to consider the operator $(J - \omega I)$.

3. Proof of One Space Dimensional Case

Proof of Theorem 2.1:

PROOF. We now begin the proof, which is composed of five steps.

Step 1. $(H(t)$ satisfies the dissipativity condition (H1).) This is done as in the proof of Theorem 2.1 of Chapter 4.

Step 2. $(H(t)$ satisfies the range condition (H2).) This is established by analogy with the proof of Theorem 2.1 in Chapter 4.

Step 3. $(H(t)$ satisfies the time-regulating condition (HA).) Let

$$g_i(x) \in C[0, 1], \quad i = 1, 2,$$

and let

$$v_1 = (I - \lambda H(t))^{-1} g_1;$$

$$v_2 = (I - \lambda H(\tau))^{-1} g_2;$$

where $\lambda > 0$ and $0 \leq t, \tau \leq T$. Then

$$(v_1 - v_2) - \lambda[\alpha(x, t, v_1')(v_1 - v_2)'' + g(x, t, v_1, v_1') - g(x, t, v_2, v_2')]$$

$$= \lambda\{\frac{\alpha(x, t, v_1') - \alpha(x, \tau, v_2')}{\alpha(x, \tau, v_2')}[H(\tau)v_2 - g(x, \tau, v_2, v_2')]$$

$$+ [g(x, t, v_2, v_2') - g(x, \tau, v_2, v_2')]\} + (g_1 - g_2);$$

$$\|v_2''\|_\infty \leq \delta_0^{-1}[\|H(\tau)v_2\|_\infty + \|g(x, \tau, v_2, v_2')\|_\infty];$$

$$(v_1 - v_2)'(0) \in [\beta_0(v_1(0)) - \beta_0(v_2(0))];$$

$$(v_1 - v_2)'(1) \in -[\beta_1(v_1(1)) - \beta_1(v_2(1))];$$

so there holds, for some function L in the condition (HA),

$$\|v_1 - v_2\|_\infty \leq \|g_1 - g_2\|_\infty + \lambda|\zeta(t) - \zeta(\tau)|L(\|v_2\|_\infty)[1 + \|H(\tau)v_2\|_\infty].$$

This proves the condition (HA), where

$$H(\tau)v_2 = \frac{v_2 - g_2}{\lambda}$$

$$= \frac{[I - \lambda H(\tau)]^{-1}g_2 - g_2}{\lambda}.$$

Here the linear growth in p at most of both $|g(x, t, z, p) - g(x, \tau, z, p)|$ and $g(x, t, z, p)$ was used, together with the interpolation inequality [1], [13, page 135]

$$\|v_2'\|_\infty \leq \epsilon\|v_2''\|_\infty + C(\epsilon)\|v_2\|_\infty$$

for each $\epsilon > 0$ and for some constant $C(\epsilon)$, where

$$\delta_0^{-1} \max_{x\in[0,1]; t\in[0,T]} |M_0(x, t, \|v_2\|_\infty)|\epsilon < 1$$

$$\text{if } \epsilon = 1/[\delta_0^{-1} \max_{x\in[0,1]; t\in[0,T]} |M_0(x, t, \|v_2\|_\infty)| + 2], \text{ for example.}$$

This is because, as in the proof of Theorem 2.1 of Chapter 4, the maximum principle applies; that is, there is an $x_0 \in (0, 1)$ such that

$$\|v_1 - v_2\|_\infty = |(v_1 - v_2)(x_0)|;$$

$$(v_1 - v_2)'(x_0) = 0;$$

$$(v_1 - v_2)(x_0)(v_1 - v_2)''(x_0) \leq 0.$$

Here $x_0 \in \{0, 1\}$ is impossible, due to the boundary conditions.

Step 4. ($H(t)$ satisfies the embedding condition (HB) of embeddedly quasi-demi-closedness.) [20] Let $t_n \in [0, T]$ converge to t, $v_n \in D(H(t_n))$ converge to v in $C[0, 1]$, and let $\|H(t_n)v_n\|_\infty$ be uniformly bounded. It will be shown that, for each η in the self-dual space $L^2(0, 1) = (L^2(0, 1))^*$, $\eta(Hv)$ exists and

$$|\eta(Hv_n) - \eta(Hv)| \longrightarrow 0.$$

Here $(C[0, 1]; \|\cdot\|_\infty)$ is continuously embedded into $L^2(0, 1); \|\cdot\|)$.

Since $\|v_n\|_\infty$ and $\|H(t_n)v_n\|_\infty$ are uniformly bounded, so is $\|v_n\|_{C^2[0,1]}$ by the interpolation inequality [1], [13, page 135]. Hence, by the Ascoli-Arzela theorem [33], a subsequence of v_n and then itself converge in $C^1[0, 1]$ to v. Also, v_n is uniformly bounded in the Hilbert space $W^{2,2}(0, 1)$, whence, by Alaoglu theorem

[36], a subsequence of v_n and then itself converge weakly to v [36]. It follows that, for each $\eta \in L^2(0,1)$,

$$|\eta(H(t_n)v_n) - \eta(H(t)v)| = |\int_0^1 (H(t_n)v_n - H(t)v)\eta \, dx|$$
$$\longrightarrow 0$$

because

$$\int_0^1 \alpha(x,t,v')(v'' - v_n'')\eta \, dx + \int_0^1 [\alpha(x,t,v') - \alpha(x,t_n,v_n')]v_n''\eta \, dx$$
$$+ \int_0^1 [g(x,t,v,v') - g(x,t_n,v_n,v_n')]\eta \, dx$$
$$\longrightarrow 0.$$

Therefore $H(t)$ satisfies the embedding condition (HB).

Step 5. ($u(t)$ for $u_0 \in \hat{D}(H(0))$ satisfies the middle equation in (1.1).) Consider the discretized equation

$$u_i - \nu H(t_i)u_i = u_{i-1},$$
$$u_i \in D(H(t_i)), \tag{3.1}$$

where $u_0 \in \hat{D}(H(0))$, $i = 1, 2, \ldots, n$, $n \in \mathbb{N}$ is large, and $\nu > 0$ is such that

$$\nu\omega < 1 \text{ and } 0 \leq t_i = i\nu \leq T.$$

Here

$$u_i = \prod_{k=1}^i [I - \nu H(t_k)]^{-1} u_0$$

exists uniquely by the range condition (H2) and the dissipativity condition (H1). For convenience, we also define

$$u_{-1} = u_0 - \nu H(0)u_0.$$

Now, for each $t \in [0,T)$, we have $t \in [t_i, t_{i+1})$ for some i, so $i = [\frac{t}{\nu}]$. It follows from Theorem 2.1 that, for each above t with the corresponding i,

$$\lim_{\nu \to 0} u_i = \lim_{\nu \to 0} \prod_{k=1}^{[\frac{t}{\nu}]} [I - \nu H(t_k)]^{-1} u_0$$
$$= \lim_{n \to \infty} \prod_{k=1}^n [I - \frac{t}{n}H(k\frac{t}{n})]^{-1} u_0$$
$$\equiv u(t)$$

exists.

On the other hand, by utilizing Proposition 4.2 in Section 4 of Chapter 2, we have that $\|u_i\|_\infty$ and $\|H(t_i)u_i\|_\infty = \|(u_i - u_{i-1})/\nu\|_\infty$ are uniformly bounded. Those, in turn, result in a bound for $\|u_i\|_{C^2[0,1]}$ by the interpolation inequality [1], [13, page 135]. Therefore it follows from Ascoli-Arzela theorem [33] that a subsequence of u_i and then itself converge in $C^1[0,1]$ to a limit, as $\nu \longrightarrow 0$. This limit equals $u(t)$ as shown above. Consequently, $u(t)$ satisfies the middle equation in (1.1), as u_i does so.

The proof is complete.   □

4. Proof of Higher Space Dimensional Case

Proof of Theorem 2.2:

PROOF. We now begin the proof, which consists of five steps.

Step 1. That $J(t)$ satisfies the dissipativity (H1) follows as in the proof of Theorem 2.2 of Chapter 4.

Step 2. The proof of Theorem 2.2 in Chapter 4 also shows that $J(t)$ satisfies the weaker range condition $(H2)'$.

Step 3. ($J(t)$ satisfies the time-regulating condition (HA).) Let

$$g_i(x) \in C^\mu(\overline{\Omega}), \quad i = 1, 2,$$

and let

$$v_1 = (I - \lambda J(t))^{-1} g_1;$$
$$v_2 = (I - \lambda J(\tau))^{-1} g_2;$$

where $\lambda > 0$ and $0 \le t, \tau \le T$. Then

$$(v_1 - v_2) - \lambda[\alpha_0(x, t, Dv_1)\triangle(v_1 - v_2) + g_0(x, t, v_1, Dv_1) - g_0(x, t, v_2, Dv_2)]$$
$$= \lambda\{\frac{\alpha_0(x, t, Dv_1) - \alpha_0(x, \tau, Dv_2)}{\alpha_0(x, \tau, Dv_2)}[J(\tau)v_2 - g_0(x, \tau, v_2, Dv_2)]$$
$$+ [g_0(x, t, v_2, Dv_2) - g_0(x, \tau, v_2, Dv_2)]\} + (g_1 - g_2), \quad x \in \Omega;$$

$$\frac{\partial(v_1 - v_2)}{\partial\hat{n}} + [\beta(x, t, v_1) - \beta(x, t, v_2)] = -[\beta(x, t, v_2) - \beta(x, \tau, v_2)], \quad x \in \partial\Omega;$$

so, there holds, for some function L in the condition (HA),

$$\|v_1 - v_2\|_\infty \le \|g_1 - g_2\|_\infty + \lambda|\zeta(t) - \zeta(\tau)|L(\|v_2\|_\infty)[1 + \|J(\tau)v_2\|_\infty]; \quad \text{or}$$
$$\|v_1 - v_2\|_\infty \le \frac{1}{\delta_1}\|N_3(x, v_2)\|_\infty|t - \tau| \le L(\|v_2\|_\infty)|t - \tau|,$$

where

$$J(\tau)v_2 = \frac{v_2 - g_2}{\lambda}$$
$$= \frac{[I - \lambda J(\tau)]^{-1} g_2 - g_2}{\lambda}.$$

Here the linear growth at most of $|g_0(x, t, z, p) - g_0(x, \tau, z, p)|$ in p was used. This is because, as in proving the dissipativity condition (H1) in Step 1, the maximum principle argument applies, that is, there is an $x_0 \in \overline{\Omega}$ such that

$$\|v_1 - v_2\|_\infty = |(v_1 - v_2)(x_0)|,$$

that, for $x_0 \in \Omega$,

$$D(v_1 - v_2)(x_0) = 0;$$
$$(v_1 - v_2)(x_0)\triangle(v_1 - v_2)(x_0) \le 0,$$

and that, for $x_0 \in \partial\Omega$,

$$\frac{\partial(v_1 - v_2)}{\partial\hat{n}}(x_0) \ge 0 \quad \text{or} \le 0 \text{ according as } (v_1 - v_2)(x_0) > 0 \text{ or} < 0.$$

Here, to derive, for some numbers $C_1(\|v_2\|_\infty)$ and $C_2(\|v_2\|_\infty)$, depending on $\|v_2\|_\infty$,

$$\|Dv_2\|_\infty \leq \|v_2\|_{C^{1+\mu}(\overline\Omega)}$$

$$\leq C_1(\|v_2\|_\infty)\|\frac{v_2-g_2}{\lambda}\|_\infty + C_2(\|v_2\|_\infty), \qquad (4.1)$$

we used the integral representation of v_2, with the Green's function $Z(x,y)$ of the second kind [29],

$$v_2 = -\int_{\partial\Omega} Z(x,y)[-\beta_2(y,\tau,v_2)]\,d\sigma_y$$

$$+ \int_\Omega Z(x,y)\alpha_0(y,\tau,Dv_2)^{-1}[\frac{v_2-g_2}{\lambda} - g_0(y,\tau,v_2,Dv_2)]\,dy.$$

Indeed, the result follows from differentiating the above v_2 for $(1+\mu)$ times, with respect to the variable x, to obtain, for some positive constants d_1 and d_2 and for some number $C_3(\|v_2\|_\infty)$ depending on $\|v_2\|_\infty$,

$$\|Dv_2\|_{C^\mu(\overline\Omega)}$$

$$\leq d_1\|\frac{v_2-g_2}{\lambda}\|_\infty$$

$$+ d_2[\max_{x\in\overline\Omega;t\in[0,T]} |M_1(x,t,\|v_2\|_\infty)|\|Dv\|_\infty + C_3(\|v_2\|_\infty)],$$

and from combining the interpolation inequality [1], [13, page 135]

$$\|Dv_2\|_\infty \leq \epsilon\|Dv_2\|_{C^\mu(\overline\Omega)} + C(\epsilon)\|v_2\|_\infty$$

for each $\epsilon > 0$ and for some constant $C(\epsilon)$. Here

$$\epsilon d_2 \max_{x\in\overline\Omega;t\in[0,T]} |M_1(x,t,\|v_2\|_\infty)| < 1$$

$$\text{if } \epsilon = 1/[d_2 \max_{x\in\overline\Omega;t\in[0,T]} |M_1(x,t,\|v_2\|_\infty)| + 2], \text{ for example.}$$

Also to be noticed are the linear growth at most of $g_0(x,t,z,p)$ in p and, for some constants b_1 and b_2,

$$|D_i Z(x,y)| \leq b_1|x-y|^{1-N};$$

$$|D_{ij} Z(x,y)| \leq b_2|x-y|^{-N};$$

$$\int_\Omega |x-y|^{N(\mu-1)}\,dx \quad \text{is finite for } 0 < \mu < 1 \text{ [13, page 159]}.$$

Step 4. $(J(t)$ satisfies the embedding condition (HB) of embeddedly quasi-demi-closedness.) [20] Let $t_n \in [0,T]$ converge to t, $v_n \in D(G(t_n))$ converge to v in $C(\overline\Omega)$, and $\|G(t_n)v_n\|_\infty$ be uniformly bounded. It will be shown that, for each η in the self-dual space $L^2(\Omega) = (L^2(\Omega))^*$, $\eta(G(t)v)$ exists and

$$|\eta(J(t_n)v_n) - \eta(J(t)v)| \longrightarrow 0.$$

Here $(C(\overline\Omega); \|\cdot\|_\infty)$ is continuously embedded into $L^2(\Omega); \|\cdot\|)$.

Since $\|v_n\|_\infty$ and $\|J(t_n)v_n\|_\infty$ are uniformly bounded, so is $\|v_n\|_{C^{1+\lambda}(\overline\Omega)}$ for any $0 < \lambda < 1$, using the proof of (4.5) in Chapter 4. (Alternatively, uniformly bounded are $\|v_n\|_{W^{2,p}(\Omega)}$ for any $p \geq 2$ and then $\|v_n\|_{C^{1+\lambda}(\overline\Omega)}$ for any $0 < \lambda < 1$, on using the L^p elliptic estimates [37] and the Sobolev embedding theorem [1, 13].) Hence, by the Ascoli-Arzela theorem [33], a subsequence of v_n and then itself converge in $C^1(\overline\Omega)$ to v. Also, v_n is uniformly bounded in the Hilbert space $W^{2,2}(\Omega)$, whence, by the Alaoglu theorem [36], a subsequence of v_n and then itself converge weakly to v [36]. It follows that, for each $\eta \in L^2(\Omega)$,

$$|\eta(G(t_n)v_n) - \eta(G(t)v)| = |\int_\Omega [G(t_n)v_n - G(t)v]\eta \, dx|$$
$$\longrightarrow 0,$$

because

$$\int_\Omega \alpha_0(x, t, Dv)(\triangle v - \triangle v_n)\eta \, dx$$
$$+ \int_\Omega [\alpha_0(x, t, Dv) - \alpha_0(x, t_n, Dv_n)]\triangle v_n \eta \, dx$$
$$+ \int_\Omega [g_0(x, t, v, Dv) - g_0(x, t_n, v_n, Dv_n)]\eta \, dx$$
$$\longrightarrow 0.$$

Therefore J satisfies the embedding condition (A3).

Step 5. ($u(t)$ for $u_0 \in \hat{D}(J(0))$ satisfies the middle equation in (1.3).) Consider the discretized equation

$$u_i - \nu J(t_i)u_i = u_{i-1},$$
$$u_i \in D(J(t_i)), \tag{4.2}$$

where $u_0 \in \hat{D}(J(0)), i = 1, 2, \ldots, n, n \in \mathbb{N}$ is large, and $\nu > 0$ is such that

$$\nu\omega < 1 \text{ and } 0 \le t_i = i\nu \le T.$$

Here

$$u_i = \prod_{k=1}^{i}[I - \nu G(t_k)]^{-1}u_0$$

exists uniquely by the range condition (H2) and the dissipativity condition (H1). For convenience, we also define

$$u_{-1} = u_0 - \nu G(0)u_0.$$

Now, for each $t \in [0, T)$, we have $t \in [t_i, t_{i+1})$ for some i, so $i = [\frac{t}{\nu}]$. It follows from Theorem 2.1 that, for each above t with the corresponding i,

$$\lim_{\nu \to 0} u_i = \lim_{\nu \to 0} \prod_{k=1}^{[\frac{t}{\nu}]}[I - \nu J(t_k)]^{-1}u_0$$
$$= \lim_{n \to \infty} \prod_{k=1}^{n}[I - \frac{t}{n}J(k\frac{t}{n})]^{-1}u_0$$
$$\equiv u(t)$$

exists.

On the other hand, by utilizing Proposition 4.2 in Section 4 of Chapter 2, we have $\|u_i\|_\infty$ and $\|J(t_i)u_i\|_\infty = \|(u_i - u_{i-1})/\nu\|_\infty$ are uniformly bounded, whence so is $\|u_i\|_{C^{1+\lambda}(\overline{\Omega})}$ for any $0 < \lambda < 1$, using the proof of (4.1). (Alternatively, those, in turn, result in a bound for $\|u_i\|_{W^{2,p}(\Omega)}$ for any $p \ge 2$, by the L^p elliptic estimates [37]. Hence, a bound exists for $\|u_i\|_{C^{1+\eta}(\overline{\Omega})}$ for any $0 < \eta < 1$, as a result of the Sobolev embedding theorem [1, 13].) Therefore it follows from Ascoli-Arzela theorem [33] that a subsequence of u_i and then itself converge in $C^{1+\mu}(\overline{\Omega})$ to a limit, as $\nu \longrightarrow 0$. This limit equals $u(t)$ as shown above. Consequently, $u(t)$ satisfies the middle equation in (1.3), as u_i does so.

The proof is complete. $\square$

CHAPTER 7

The Associated Elliptic Equations

1. Introduction

In this chapter, the solutions for the associated elliptic equations with solving the parabolic problems in Chapters 3, 4, 5, and 6 will be further shown to be explicit functions of the solution ϕ to the elliptic equation

$$-\triangle v(y) = h, \quad y \in \Omega;$$

$$\frac{\partial v}{\partial \hat{n}} + v = 0, \quad y \in \partial\Omega.$$

Here for the dimension of the space variable y equal to 2 or 3, the ϕ can be computed numerically and efficiently by the boundary element methods. As a consequence, a solution for a parabolic, initial-boundary value problem in the previous chapters might be computed numerically.

Thus we recall the following.

In Chapter 6, the nonlinear, non-autonomous, parabolic initial-boundary value problem was considered

$$u_t(x,t) = \alpha_0(x,t,Du)\triangle u(x,t) + g_0(x,t,u,Du),$$
$$(x,t) \in \Omega \times (0,T);$$

$$\frac{\partial}{\partial \hat{n}} u(x,t) + \beta(x,t,u) = 0, \quad x \in \partial\Omega; \qquad (1.1)$$

$$u(x,0) = u_0(x).$$

It was written as the nonlinear evolution equation

$$\frac{d}{dt} u(t) = J(t)u(t), \quad t \in (0,T); \qquad (1.2)$$

$$u(0) = u_0,$$

where the time-dependent, nonlinear operator

$$J(t) : D(J(t)) \subset C(\overline{\Omega}) \longrightarrow C(\overline{\Omega})$$

was defined by

$$J(t)v = \alpha_0(x,t,Dv)\triangle v + g_0(x,t,v,Dv);$$

$$v \in D(J(t)) \equiv \{w \in C^{2+\mu}(\overline{\Omega}) : \frac{\partial}{\partial \hat{n}} w + \beta(x,t,w) = 0 \quad \text{on } \partial\Omega\}.$$

It was then proved that, for $u_0 \in \hat{D}(J(0))$, the quantity

$$u(t) = \lim_{n\to\infty} \prod_{i=1}^{n} [I - \frac{t}{n} J(i\frac{t}{n})]^{-1} u_0$$

$$= \lim_{\nu \to 0} \prod_{i=1}^{[\frac{t}{\nu}]} [I - \nu J(i\nu)]^{-1} u_0$$

is a strong solution for equation (1.2) or, equivalently, equation (1.1). Here the quantity $[I - \frac{t}{n}J(i\frac{t}{n})]^{-1}u_0$ is the unique solution to the elliptic equation

$$v - \frac{t}{n}J(i\frac{t}{n})v = u_0,$$

that is associated with the nonlinear parabolic equation (1.1).

In this chapter, it will be further shown that the quantity, for $h \in C^\mu(\overline{\Omega})$ where $0 < \mu < 1$,

$$[I - \frac{t}{n}J(i\frac{t}{n})]^{-1}h, \quad i = 1, 2, \ldots, n$$

is the limit of a sequence where each term in the sequence is an explicit function of the solution ϕ to the elliptic equation

$$\begin{aligned} -\Delta v(y) &= h, \quad y \in \Omega; \\ \frac{\partial v}{\partial \hat{n}} + v &= 0, \quad y \in \partial\Omega. \end{aligned} \tag{1.3}$$

Here for the dimension of the space variable y equal to 2 or 3, the ϕ can be computed numerically and efficiently by the boundary element methods [11, 34]. Thus, a solution for a parabolic, initial-boundary value problem in the previous chapters might be computed numerically.

The same thing will be done to the linear, non-autonomous, parabolic initial-boundary value problem in Chapter 5

$$u_t(x,t) = \sum_{i,j=1}^{N} a_{ij}(x,t)D_{ij}u(x,t) + \sum_{i=1}^{N} b_i(x)D_i u(x,t)$$
$$+ c(x)u(x,t) + f_0(x,t), \quad (x,t) \in \Omega \times (0,T); \tag{1.4}$$
$$\frac{\partial}{\partial \hat{n}}u(x,t) + \beta_2(x,t)u(x,t) = 0, \quad x \in \partial\Omega;$$
$$u(x,0) = u_0(x),$$

where $\sum_{i,j=1}^{N} a_{ij}(x,t)D_{ij}u(x,t) = a_0(x,t)\Delta u(x,t)$. To (1.4), the corresponding evolution equation there in Chapter 5 was

$$\begin{aligned} \frac{d}{dt}u(t) &= G(t)u(t), \quad 0 < t < T; \\ u(0) &= u_0, \end{aligned} \tag{1.5}$$

in which the time-dependent operator

$$G(t) : D(G(t)) \subset C(\overline{\Omega}) \longrightarrow C(\overline{\Omega})$$

was defined by

$$G(t)v = \sum_{i,j=1}^{N} a_{ij}(x,t)D_{ij}v + \sum_{i=1}^{N} b_i(x,t)D_i v + c(x,t)v + f_0(x,t);$$

$$v \in D(G(t)) \equiv \{w \in C^{2+\mu}(\overline{\Omega}) : \frac{\partial}{\partial \hat{n}}w + \beta_2(x,t)w = 0 \quad \text{on } \partial\Omega\}.$$

It was shown there in Chapter 5 that, for $u_0 \in \hat{D}(G(0))$, the quantity

$$u(t) = \lim_{n \to \infty} \prod_{i=1}^{n} [I - \frac{t}{n} G(i\frac{t}{n})]^{-1} u_0$$

$$= \lim_{\nu \to 0} \prod_{i=1}^{[\frac{t}{\nu}]} [I - \nu G(i\nu)]^{-1} u_0$$

is a strong solution for equation (1.5) or, equivalently, equation (1.4).

The rest of this chapter is organized as follows. Section 2 states the main results, and Sections 3 and 4 prove the main results.

The material of this chapter is taken from [**25**].

2. Main Results

Here we first consider the linear, time-dependent operator $G(t)$ in (1.5), where $\sum_{i,j=1}^{N} a_{ij}(x,t)D_{ij}u(x,t)$ is not necessarily equal to $a_0(x,t)\triangle u(x,t)$. In this case, we additionally assume that $a_{ij}(x,t) = a_{ji}(x,t)$ is in $C^\mu(\overline{\Omega})$ for all $t \in [0,T]$, and satisfies, for $\xi \in \mathbb{R}^N, x \in \overline{\Omega}$, and $t \in [0,t]$,

$$\lambda_{min}|\xi|^2 \leq \sum_{i,j=1}^{N} a_{ij}\xi_i\xi_j \leq \lambda_{max}|\xi|^2$$

for some positive constants λ_{min} and λ_{max}.

THEOREM 2.1. *For $h \in C^\mu(\overline{\Omega})$, the solution u to the equation*

$$[I - \epsilon G(t)]u = h \tag{2.1}$$

where $0 \leq t \leq T$ and $\epsilon > 0$, is the limit of a sequence where each term in the sequence is an explicit function of the solution ϕ to the elliptic equation (1.3).

We next consider the case with the nonlinear, time-dependent operator $J(t)$ in (1.2).

THEOREM 2.2. *For $h \in C^\mu(\overline{\Omega})$, the solution u to the equation*

$$[I - \epsilon J(t)]u = h \tag{2.2}$$

where $0 \leq t \leq T$ and $\epsilon > 0$, is the limit of a sequence where each term in the sequence is an explicit function of the solution ϕ to the elliptic equation (1.3). Here $\beta(x,t,0) \equiv 0$ is assumed additionally.

3. Proof of Linear Case

Proof of Theorem 2.1:

PROOF. Solvability of the equation (2.1) follows from [**13**, Pages 128-130], where the method of continuity [**13**, Page 75] is used. By writing out fully how the method of continuity is used, it will be seen that the solution u is the limit of a sequence where each term in the sequence is an explicit function of the solution ϕ to the elliptic equation (1.3).

To this end, set

$$U_1 = C^{2+\mu}(\overline{\Omega}); \quad U_2 = C^{\mu}(\overline{\Omega}) \times C^{1+\mu}(\partial\Omega);$$

$$L_\tau u = \tau[u - \epsilon G(t)u] + (1 - \tau)(-\triangle u) \text{ in } \Omega;$$

$$N_\tau u = \tau[\frac{\partial u}{\partial \hat{n}} + \beta(x, t)u] + (1 - \tau)(\frac{\partial u}{\partial \hat{n}} + u) \text{ on } \partial\Omega;$$

where $0 \leq \tau \leq 1$. Define the linear operator $\pounds_\tau : U_1 \longrightarrow U_2$ by

$$\pounds_\tau u = (L_\tau u, N_\tau u)$$

for $u \in U_1$, and assume that $\pounds_s$ is onto for some $s \in [0, 1]$.

It follows from [**13**, Pages 128-130] in which used are the maximum principle and the Schauder global estimate [**13**] that

$$\|u\|_{U_1} \leq C\|\pounds_\tau u\|_{U_2}, \tag{3.1}$$

where the constant C is independent of τ. This implies that $\pounds_s$ is one to one, so $\pounds_s^{-1}$ exists. By making use of $\pounds_s^{-1}$, the equation, for $w_0 \in U_2$ given,

$$\pounds_\tau u = w_0$$

is equivalent to the equation

$$u = \pounds_s^{-1} w_0 + (\tau - s)\pounds_s^{-1}(\pounds_0 - \pounds_1)u,$$

from which a linear map

$$S : U_1 \longrightarrow U_1,$$

$$Su = S_s u \equiv \pounds_s^{-1} w_0 + (\tau - s)\pounds_s^{-1}(\pounds_0 - \pounds_1)u$$

is defined. The unique fixed point u of $S = S_s$ will be related to the solution of (2.1).

By choosing $\tau \in [0, 1]$ such that

$$|s - \tau| < \delta \equiv [C(\|\pounds_0\|_{U_1 \to U_2} + \|\pounds_1\|_{U_1 \to U_2})]^{-1}, \tag{3.2}$$

it follows that $S = S_s$ is a strict contraction map. Therefore S has a unique fixed point w, and the w can be represented by

$$\lim_{n \to \infty} S^n 0 = \lim_{n \to \infty} (S_s)^n 0$$

because of $0 \in U_1$. Thus $\pounds_\tau$ is onto for $|\tau - s| < \delta$.

It follows that, by dividing $[0, 1]$ into subintervals of length less than δ and repeating the above arguments in a finite number of times, $\pounds_\tau$ becomes onto for all $\tau \in [0, 1]$, provided that it is onto for some $\tau \in [0, 1]$. Since $\pounds_0$ is onto by the potential theory [**13**, Page 130], we have that $\pounds_1$ is also onto. Therefore, for $w_0 = (h, 0)$, the equation

$$\pounds_1 u = w_0$$

has a unique solution u, and the u is the sought solution to (2.1). Here it is to be observed that $\phi \equiv \pounds_0^{-1}(h, 0)$ is the unique solution $\pounds_0^{-1}(h, \varphi)$ to the elliptic equation (3.3) with $\varphi \equiv 0$:

$$- \triangle v = h, \quad x \in \Omega,$$

$$\frac{\partial v}{\partial \hat{n}} + v(x) = \varphi \text{ on } \partial\Omega, \tag{3.3}$$

and that

$$S0 = S_0 0 = \pounds_0^{-1}(h, 0),$$

$$S^2 0 = (S_0)^2 0 = \pounds_0^{-1}(h, 0) + \pounds_0^{-1}[|\tau - 0|(\pounds_0 - \pounds_1)\pounds_0^{-1}(h, 0)],$$

$$\vdots$$

The proof is complete.   □

Remark.

- The solution u is eventually represented by

$$u(x) = \pounds_0^{-1} H((h, 0)),$$

where $H((h, 0))$ is a convergent series in which each term is basically obtained by, repeatedly, applying the linear operator $(\pounds_0 - \pounds_1)\pounds_0^{-1}$ to $(h, 0)$ for a certain number of times.
- The quantity $\pounds_0^{-1}(h, \varphi)$, for each $(h, \varphi) \in U_2$ given, can be computed numerically and efficiently by the boundary element methods [11, 34], if the dimension of the space variable x equals 2 or 3.
- The constant C above in (3.1) and (3.2) depends on $n, \mu, \lambda_{min}, \Omega$, and on the coefficient functions $a_{ij}(x, t), b_i(x, t), c(x, t), \beta(x, t)$, and is not known explicitly [13]. Therefore, the corresponding δ cannot be determined in advance. Thus, when dealing with the elliptic equation (2.1) in Theorem 2.1 numerically, it is more possible, by choosing $\tau \in [0, 1]$ such that $|s - \tau|$ is smaller, that the sequence $S^n 0$ will converge, for which

$$|s - \tau| < \delta$$

occurs.

4. Proof of the Nonlinear Case

Proof of Theorem 2.2:

PROOF. The equation (2.2) has been solved in Chapter 6, but here the proof will be based on the contraction mapping theorem as in the proof of Theorem 2.1. To this end, set

$$U_1 = C^{2+\mu}(\overline{\Omega});$$

$$U_2 = C^{\mu}(\overline{\Omega}) \times C^{1+\mu}(\partial\Omega);$$

$$L_\tau u = \tau[u - \epsilon J(t)u] + (1 - \tau)(u - \triangle u), \quad x \in \Omega;$$

$$N_\tau u = \tau[\frac{\partial u}{\partial \hat{n}} + \beta(x, t, u)] + (1 - \tau)(\frac{\partial u}{\partial \hat{n}} + u) \text{ on } \partial\Omega;$$

where $0 \leq \tau \leq 1$. Define the nonlinear operator $\pounds_\tau : U_1 \longrightarrow U_2$ by

$$\pounds_\tau u = (L_\tau u, N_\tau u)$$

for $u \in U_1$, and assume that $\pounds_s$ is onto for some $s \in [0, 1]$.

As in proving that $J(t)$ satisfies the dissipativity (H1) where the maximum principle was used, $\pounds_s$ is one to one, so $\pounds_s^{-1}$ exists. By making use of $\pounds_s^{-1}$, the equation, for $w_0 \in U_2$ given,

$$\pounds_\tau u = w_0$$

is equivalent to the equation

$$u = \pounds_s^{-1}[w_0 + (\tau - s)(\pounds_0 - \pounds_1)u],$$

from which a nonlinear map

$$S : U_1 \longrightarrow U_1,$$
$$Su = S_s u \equiv \pounds_s^{-1}[w_0 + (\tau - s)(\pounds_0 - \pounds_1)u] \quad \text{for } u \in U_1$$

is defined. The unique fixed point of $S = S_s$ will be related to the solution of (2.2). By restricting $S = S_s$ to the closed ball of the Banach space U_1,

$$B_{s,r,w_0} \equiv \{u \in U_1 : \|u - \pounds_s^{-1}w_0\|_{C^{2+\mu}} \leq r > 0\},$$

and choosing small enough $|\tau - s|$, we will show that $S = S_s$ leaves B_{s,r,w_0} invariant. This will be done by the following Steps 1 to 4.

Step 1. It follows as in Chapter 6 that for $\pounds_\tau v = (f, \chi)$,

$$
\begin{aligned}
&\|v\|_\infty \leq k_{\{\|f\|_\infty, \|\chi\|_{C(\partial\Omega)}\}}; \\
&\|Dv\|_{C^\mu} \leq k_{\{\|v\|_\infty\}}\|Dv\|_\infty + k_{\{\|v\|_\infty, \|f\|_\infty, \|\chi\|_{C(\partial\Omega)}\}}; \\
&\|v\|_{C^{1+\mu}} \leq k_{\{\|\chi\|_{C(\partial\Omega)}, \|f\|_\infty\}}; \\
&\|v\|_{C^{2+\mu}} \leq K\|\pounds_\tau v\|_{U_2} = K\|\pounds_\tau v\|_{C^\mu(\overline{\Omega}) \times C^{1+\mu}(\partial\Omega)}.
\end{aligned}
\tag{4.1}
$$

Here $k_{\{\|f\|_\infty\}}$ is a constant depending on $\|f\|_\infty$, and similar meaning is defined for other constants k's; further, K is independent of τ, and while it depends on $N, \lambda_{min}, \mu, \Omega$, and on the $C^{1+\mu}(\overline{\Omega})$ norm of the coefficient functions

$$\alpha_0(x, t, Dv), g_0(x, t, v, Dv), \beta(x, t, v),$$

the K has incorporated its dependence on $\|v\|_{C^{1+\mu}}$ into $\|\pounds_\tau v\|_{U_2}$, by using the third equation in (4.1).

Step 2. It is readily seen that, for $v \in C^{2+\mu}(\overline{\Omega})$ with $\|v\|_{C^{2+\mu}} \leq R > 0$, we have

$$\|\pounds_\tau v\|_{U_2} \leq k_{\{R\}}\|v\|_{C^{2+\mu}},
\tag{4.2}$$

where $k_{\{R\}}$ is independent of τ.

Step 3. It will be shown that, if

$$\|u\|_{C^{2+\mu}} \leq R, \quad \|v\|_{C^{2+\mu}} \leq R > 0,$$

then

$$\|\pounds_\tau u - \pounds_\tau v\|_{U_2} \leq k_{\{R\}}\|u - v\|_{C^{2+\mu}}.
\tag{4.3}$$

It will be also shown that, if

$$\pounds_\tau u = (f, \chi_1), \quad \pounds_\tau v = (w, \chi_2),$$

then

$$
\begin{aligned}
\|u - v\|_{C^{2+\mu}} &\leq k_{\{\|\pounds_\tau u\|_{U_2}, \|\pounds_\tau v\|_{U_2}\}}[\|f - w\|_{C^\mu} + \|\chi_1 - \chi_2\|_{C^{1+\mu}}] \\
&= k_{\{\|\pounds_\tau u\|_{U_2}, \|\pounds_\tau v\|_{U_2}\}}\|\pounds_\tau u - \pounds_\tau v\|_{U_2}.
\end{aligned}
\tag{4.4}
$$

Here $K_{\{R\}}$ and $K_{\{\|\pounds_\tau u\|_{U_2}, \|\pounds_\tau v\|_{U_2}\}}$ are independent of τ.

Using the mean value theorem, we have that

$$f - w = L_\tau u - L_\tau v$$
$$= (u - v) - (1 - \tau)\triangle(u - v) - \tau\epsilon[\alpha\triangle(u - v)$$
$$+ \alpha_p(x, t, p_1)(Du - Dv)\triangle v + g_p(x, t, u, p_2)(Du - Dv)$$
$$+ g_z(x, t, z_1, Dv)(u - v)], \quad x \in \Omega;$$

$$\frac{\partial(u - v)}{\partial\hat{n}} + [\beta(x, t, u) - \beta(x, t, v)] = \chi_1 - \chi_2 \quad \text{on } \partial\Omega;$$

where p_1, p_2 are some functions between Du and Dv, and z_1 is some function between u and v.

It follows as in (4.2) that

$$\|\mathcal{L}_\tau u - \mathcal{L}_\tau v\|_{U_2} \leq k_{\{R\}}\|u - v\|_{C^{2+\mu}},$$

which is one desired estimate.

On the other hand, the maximum principle yields

$$\|u - v\|_\infty \leq k_{\{\|f - w\|_\infty, \|\chi_1 - \chi_2\|_\infty\}},$$

and the fourth equation in (4.1) delivers

$$\|u\|_{C^{2+\mu}} \leq K\|\mathcal{L}_\tau u\|_{U_2};$$
$$\|v\|_{C^{2+\mu}} \leq K\|\mathcal{L}_\tau v\|_{U_2}.$$

Thus, it follows from the Schauder global estimate [13] that

$$\|u - v\|_{C^{2+\mu}} \leq k_{\{\|\mathcal{L}_\tau u\|_{U_2}, \|\mathcal{L}_\tau\|_{U_2}\}}\|\mathcal{L}_\tau u - \mathcal{L}_\tau v\|_{U_2},$$

the other desired estimate.

Step 4. Consequently, for $u \in B_{s,r,w_0}$, we deduce that, by the fourth equation in (4.1),

$$\|u\|_{C^{2+\mu}} \leq r + \|\mathcal{L}_s^{-1}w_0\|_{C^{2+\mu}} \leq r + K\|w_0\|_{U_2}$$
$$\equiv R_{\{r, \|w_0\|_{U_2}\}}, \tag{4.5}$$

and that

$$\|Su - \mathcal{L}_s^{-1}w_0\|_{C^{2+\mu}}$$
$$\leq k_{\{\|w_0\|_{U_2}, \|w_0 + (\tau - s)(\mathcal{L}_0 - \mathcal{L}_1)u\|_{U_2}\}}\|(\tau - s)(\mathcal{L}_0 - \mathcal{L}_1)u\|_{U_2} \quad \text{by (4.4)}$$
$$\leq |\tau - s|k_{\{\|w_0\|_{U_2}, R_{\{r, \|w_0\|_{U_2}\}}\}} \quad \text{by (4.2) and (4.5)}.$$

Here the constant $k_{\{\|w_0\|_{U_2}, R_{\{r, \|w_0\|_{U_2}\}}\}}$ when w_0 given and r chosen, is independent of τ and s. Hence, by choosing some sufficiently small $\delta_1 > 0$, there results

$$S = S_s : B_{s,r,w_0} \subset U_1 \longrightarrow B_{s,r,w_0} \subset U_1$$

for $|\tau - s| < \delta_1$; that is, B_{s,r,w_0} is left invariant by $S = S_s$.

Next, it will be shown that, for small $|\tau - s|$, $S = S_s$ is a strict contraction on B_{s,r,w_0}, from which $S = S_s$ has a unique fixed point. Because, for $u, v \in B_{s,r,w_0}$,

$$\|u\|_{C^{2+\mu}} \leq R_{\{r, \|w_0\|_{U_2}\}}, \quad \|v\|_{C^{2+\mu}} \leq R_{\{r, \|w_0\|_{U_2}\}} \quad \text{by (4.5)},$$

it follows that, by (4.2),

$$\|w_0 + (\tau - s)(\mathcal{L}_0 - \mathcal{L}_1)u\|_{U_2} \leq k_{\{\|w_0\|_{U_2}, R_{\{r, \|w_0\|_{U_2}\}}\}};$$
$$\|w_0 + (\tau - s)(\mathcal{L}_0 - \mathcal{L}_1)v\|_{U_2} \leq k_{\{\|w_0\|_{U_2}, R_{\{r, \|w_0\|_{U_2}\}}\}}; \tag{4.6}$$

and that, by (4.3),

$$\|(\tau - s)[(\pounds_0 - \pounds_1)u - (\pounds_0 - \pounds_1)v]\|_{U_2}$$
$$\leq |\tau - s| k_{\{R_{\{r, \|w_0\|_{U_2}\}}\}} \|u - v\|_{C^{2+\mu}}. \tag{4.7}$$

Therefore, on account of (4.4), (4.6), and (4.7), we obtain

$$\|Su - Sv\|_{C^{2+\mu}}$$
$$\leq |\tau - s| k_{\{R_{\{r, \|w_0\|_{U_2}\}}, \|w_0\|_{U_2}\}} k_{\{R_{\{r, \|w_0\|_{U_2}\}}\}} \|u - v\|_{C^{2+\mu}}.$$

Here the constant $k_{\{R_{\{r, \|w_0\|_{U_2}\}}, \|w_0\|_{U_2}\}} k_{\{R_{\{r, \|w_0\|_{U_2}\}}\}}$ when w_0 given and r chosen, is independent of τ and s. Hence, by choosing some sufficiently small $\delta_2 > 0$, it follows that

$$S = S_s : B_{s,r,w_0} \longrightarrow B_{s,r,w_0}$$

is a strict contraction for

$$|\tau - s| < \delta_2 \leq \delta_1.$$

Furthermore, the unique fixed point w of $S = S_s$ can be represented by

$$\lim_{n \to \infty} S^n 0 = \lim_{n \to \infty} (S_s)^n 0$$

if $\beta(x, t, 0) \equiv 0$ and if $r = r_{\{K\|w_0\|_{U_2}\}}$ is chosen such that

$$r = r_{\{K\|w_0\|_{U_2}\}} \geq K\|w_0\|_{U_2}$$
$$\geq \|\pounds_s^{-1} w_0\|_{C^{2+\mu}} \quad \text{by (4.5);} \tag{4.8}$$

this is because 0 belongs to B_{s,r,w_0} in this case. Thus $\pounds_\tau$ is onto for $|\tau - s| < \delta_2$.

It follows that, by dividing $[0, 1]$ into subintervals of length less than δ_2 and repeating the above arguments in a finite number of times, $\pounds_\tau$ becomes onto for all $\tau \in [0, 1]$, provided that it is onto for some $\tau \in [0, 1]$. Since $\pounds_0$ is onto by linear elliptic equations theory [13], we have that $\pounds_1$ is also onto. Therefore, the equation, for $w_0 = (h, 0)$,

$$\pounds_1 u = w_0$$

has a unique solution u, and the u is the sought solution to (2.2).

Here it is to be observed that $\psi \equiv \pounds_0^{-1}(h, 0)$ is the unique solution to the elliptic equation

$$v - \Delta v = h, \quad x \in \Omega,$$
$$\frac{\partial v}{\partial \hat{n}} + v(x) = 0 \text{ on } \partial\Omega,$$

and that, by Theorem 2.1, the ψ is the limit of a sequence where each term in the sequence is an explicit function of the solution ϕ to the elliptic equation (1.3).

It is also to be observed that

$$S0 = S_0 0 = \pounds_0^{-1}(h, 0),$$
$$S^2 0 = (S_0)^2 0 = \pounds_0^{-1}[(h, 0) + |\tau - 0|(\pounds_0 - \pounds_1)\pounds_0^{-1}(h, 0)],$$

$$\vdots,$$

where $(\pounds_0 - \pounds_1)\pounds_0^{-1}$ is a nonlinear operator.

The proof is complete.   □

Remark.

- The constants $k_{\{R_{\{r,\|w_0\|_{U_2}\}}\}}$ and $k_{\{R_{\{r,\|w_0\|_{U_2}\}},\|w_0\|_{U_2}\}}k_{\{R_{\{r,\|w_0\|_{U_2}\}}\}}$, when w_0 is given and when r is chosen and conditioned by (4.8), is not known explicitly, so the corresponding δ_2 cannot be determined in advance. Hence, when dealing with the elliptic equation (2.2) in Theorem 2.2 numerically, it is more possible, by choosing $\tau \in [0,1]$ such that $|\tau - s|$ is smaller, that the sequence $S^n 0$ will converge, for which

$$|\tau - s| < \delta_2 \le \delta_1$$

occurs.

CHAPTER 8

Existence Theorems for Evolution Equations (II)

1. Introduction

In Chapter 6, the two initial-boundary value problems for parabolic, partial differential equations with the Robin boundary conditions were solved by applying the results in Chapter 2:

$$u_t(x,t) = \alpha(x,t,u_x)u_{xx}(x,t) + g(x,t,u,u_x),$$
$$(x,t) \in (0,1) \times (0,T);$$
$$u_x(0,t) \in \beta_0(u(0,t)), \quad u_x(1,t) \in -\beta_1(u(1,t));$$
$$u(x,0) = u_0(x),$$

and

$$u_t(x,t) = \alpha_0(x,t,Du)\triangle u(x,t) + g_0(x,t,u,Du),$$
$$(x,t) \in \Omega \times (0,T);$$
$$\frac{\partial}{\partial \hat{n}}u(x,t) + \beta(x,t,u) = 0, \quad x \in \partial\Omega;$$
$$u(x,0) = u_0(x).$$

However, the results in Chapter 2 become inapplicable, if the coefficient functions in the leading terms, $\alpha = \alpha(x,t,u,u_x)$ and $\alpha_0 = \alpha_0(x,t,u,Du)$, have the u dependence. This is due to the failure of the dissipativity ccondition (H1) in Chapter 2, as is checked by the maximum principle arguments. But this problem will be resolved here, as the following describes it.

In this chapter, we will continue the study in Chapter 2 by changing its dissipativity condition (H1), range condition (H2), and time-regulating condition (HA) or $(HA)'$. The results obtained will be applied to solve more general, nonlinear parabolic partial differential equations in Chapter 9, where the coefficient function in the leading term has the u dependence additionally. Because of this u dependence, the dissipativity condition (H1) will fail and the results obtained in Chapter 2 cannot be applied. The failure of (H1) is readily checked by the familiar maximum principle arguments in Sections 3 and 4 in Chapter 9 (or in Sections 3 and 4 in Chapter 6).

As is the case with Chapter 2, let $(X, \|\cdot\|)$ be a real Banach space with the norm $\|\cdot\|$, and let $T > 0$ be a real constant. Consider the nonlinear evolution equation

$$\frac{du(t)}{dt} \in A(t)u(t), \quad 0 \leq s < t < T,$$
$$u(s) = u_0,$$
(1.1)

where
$$A(t) : D(A(t)) \subset X \longrightarrow X$$
is a nonlinear, time-dependent, and multi-valued operator.

Equation (1.1) will be solved under the set of hypotheses, namely, the non-dissipativity condition (H3), the weaker range condition $(H2)'$, and the time-regulating condition (HC) or $(HC)'$.

$(H2)'$ The range of $(I - \lambda A(t))$, denoted by E, is independent of t and contains $D(A(t))$ for all $t \in [0,T]$ and for small $0 < \lambda < \lambda_0$, where λ_0 is some positive number.

(H3) For all $0 \le t \le T$ and all $0 < \lambda < \lambda_0$, the λ_0 in $(H2)'$, $A(t)$ satisfies the non-dissipativity condition. Namely, there are some number δ_0 with $\lambda_0 \delta_0 < 1$ and some nonnegative numbers a and $\tilde{a}$, and there are some function L_0 and some nonnegative function L_1 on $[0,\infty) \times [0,\infty)$, such that, for all $u, v \in E$, the following hold:

$$\|\tilde{u}\| \le [(1 - \lambda \delta_0)^{-1}(\|u\| + \lambda a) \quad \text{or} \quad \tilde{a}],$$
$$\text{where } \tilde{u} \in J_\lambda(t)u \equiv (I - \lambda A(t))^{-1}u;$$
$$\|\tilde{u} - \tilde{v}\| \le [\|u - v\| + \lambda L_0(\|\tilde{u}\|, \|\tilde{v}\|)\|\tilde{u} - \tilde{v}\|$$
$$+ \lambda L_1(\|\tilde{u}\|, \|\tilde{v}\|)\frac{\|\tilde{v} - v\|}{\lambda}\|\tilde{u} - \tilde{v}\|],$$
$$\text{where } \tilde{u} \in J_\lambda(t)u \text{ and } \tilde{v} \in J_\lambda(t)v.$$

Here $L_k, k = 0, 1$ are bounded on bounded subsets of $[0,\infty) \times [0,\infty)$, and, for $\tau, s_1, s_2 \ge 0$, the expression
$$\tau \le [s_1 \text{ or } s_2]$$
means
$$\tau \le s_1 \text{ or } \tau \le s_2.$$
In fact, it will be defined for later use that
$$\tau \le [s_1 \text{ or } s_2 \text{ or } \dots \text{ or } s_n]$$
means
$$\tau \le s_1 \text{ or } \tau \le s_2 \text{ or } \dots \text{ or } \tau \le s_n,$$
and that
$$\tau \le [s_1 \text{ and } s_2 \text{ and } \dots \text{ and } s_n]$$
means
$$\tau \le s_1 \text{ and } \tau \le s_2 \text{ and } \dots \text{ and } \tau \le s_n.$$
Here $\tau, s_i \ge 0, i = 1, 2, \dots, n \in \mathbb{N}$.

To be noticed is that the dissipativity condition (H1) implies the non-dissipativity condition (H3) here, if $0 \in D(A(t))$ for all t and
$$\zeta \equiv \sup_{v_0 \in A(t)0; t \in [0,T]} \|v_0\|$$
is finite. This follows because then $a = \zeta, L_1 \equiv 0$, and $L_0 \equiv \delta_0 = \omega$, where the constant ω is from (H1).

(HC) There are two continuous functions $f, g : [0, T] \longrightarrow \mathbb{R}$, of bounded variation, and there are one function M_1 and four nonnegative functions $M_k, k = 0, 2, 3, 4$, on $[0, \infty) \times [0, \infty)$ with $M_k(s_1, s_2)$ bounded for bounded s_1, s_2, such that, for each $0 < \lambda < \lambda_0$, the λ_0 in $(H2)'$, we have

$$S_1(\lambda) \cup S_2(\lambda)$$
$$\equiv \{\tilde{x} - \tilde{y} : \tilde{x} \in J_\lambda(t)x, \tilde{y} \in J_\lambda(\tau)y; 0 \le t, \tau \le T; x, y \in E\}.$$

Here $S_1(\lambda)$ denotes the set:

$$\{\tilde{x} - \tilde{y} : \tilde{x} \in J_\lambda(t)x, \tilde{y} \in J_\lambda(\tau)y; 0 \le t, \tau \le T; x, y \in E;$$
$$\|\tilde{x} - \tilde{y}\| \le M_0(\|\tilde{x}\|, \|\tilde{y}\|)|t - \tau|\},$$

while $S_2(\lambda)$ denotes the set:

$$\{\tilde{x} - \tilde{y} : \tilde{x} \in J_\lambda(t)x, \tilde{y} \in J_\lambda(\tau)y; 0 \le t, \tau \le T; x, y \in E;$$
$$\|\tilde{x} - \tilde{y}\| \le [\|x - y\| + \lambda M_1(\|\tilde{x}\|, \|\tilde{y}\|)\|\tilde{x} - \tilde{y}\|$$
$$+ \lambda M_2(\|\tilde{x}\|, \|\tilde{y}\|)\frac{\|\tilde{y} - y\|}{\lambda}\|\tilde{x} - \tilde{y}\|$$
$$+ \lambda|f(t) - f(\tau)|M_3(\|\tilde{x}\|, \|\tilde{y}\|)\frac{\|\tilde{y} - y\|}{\lambda}$$
$$+ \lambda|g(t) - g(\tau)|M_4(\|\tilde{x}\|, \|\tilde{y}\|)]\}.$$

Observed is that (HC) is reduced to the second condition in (H3) when $S_1(\lambda) = \emptyset$ and $t = \tau$.

(HC)' There are two continuous functions $f, g : [0, T] \longrightarrow \mathbb{R}$, of bounded variation, and there are one function $\tilde{M}_1$ and four nonnegative functions $\tilde{M}_k, k = 0, 2, , 3, 4$, on $[0, \infty)^2 \times [0, \infty)^2$ with $\tilde{M}_k(s_1, s_2, s_3, s_4)$ bounded for bounded $s_i, i = 1, 2, 3, 4$, such that, for each $0 < \lambda < \lambda_0$, the λ_0 in $(H2)'$, we have

$$S_1(\lambda) \cup S_2(\lambda)$$
$$\equiv \{\tilde{x} - \tilde{y} : \tilde{x} \in J_\lambda(t)x, \tilde{y} \in J_\lambda(\tau)y; 0 \le t, \tau \le T; x, y \in E\}.$$

Here $S_1(\lambda)$ denotes the set:

$$\{\tilde{x} - \tilde{y} : \tilde{x} \in J_\lambda(t)x, \tilde{y} \in J_\lambda(\tau)y; 0 \le t, \tau \le T; x, y \in E;$$
$$\|\tilde{x} - \tilde{y}\| \le \tilde{M}_0(\|\tilde{x}\|, \|x\|, \|\tilde{y}\|, \|y\|)|t - \tau|\},$$

while $S_2(\lambda)$ denotes the set:

$$\{\tilde{x} - \tilde{y} : \tilde{x} \in J_\lambda(t)x, \tilde{y} \in J_\lambda(\tau)y; 0 \le t, \tau \le T; x, y \in E;$$
$$\|\tilde{x} - \tilde{y}\| \le [\|x - y\| + \lambda \tilde{M}_1(\|\tilde{x}\|, \|x\|, \|\tilde{y}\|, \|y\|)\|\tilde{x} - \tilde{y}\|$$
$$+ \lambda \tilde{M}_2(\|\tilde{x}\|, \|x\|, \|\tilde{y}\|, \|y\|)\frac{\|\tilde{y} - y\|}{\lambda}\|\tilde{x} - \tilde{y}\|$$
$$+ \lambda|f(t) - f(\tau)|\tilde{M}_3(\|\tilde{x}\|, \|x\|, \|\tilde{y}\|, \|y\|)\frac{\|\tilde{y} - y\|}{\lambda}$$
$$+ \lambda|g(t) - g(\tau)|\tilde{M}_4(\|\tilde{x}\|, \|x\|, \|\tilde{y}\|, \|y\|)]\}.$$

Again, $(HC)'$ becomes essentially the second condition in (H3), as is the case with (HC), when $S_1(\lambda) = \emptyset$ and $t = \tau$.

The purpose of this chapter is to show, in a way similar to Chapter 2, that, with $(H2)'$, (H3), and (HC) or $(HC)'$ assumed, the quantity, for $x \in D(A(s))$ and for large enough $n \in \mathbb{N}$,

$$\prod_{i=1}^{n} J_{\frac{t-s}{n}}\left(s + i\frac{t-s}{n}\right)x$$

is single-valued and its limit, as $n \longrightarrow \infty$,

$$U(t,s)x \equiv \lim_{n\to\infty} \prod_{i=1}^{n} J_{\frac{t-s}{n}}\left(s + i\frac{t-s}{n}\right)x$$

exists, provided that certain condition of smallness is satisfied. Again, this limit $U(t,s)x$ for $x = u_0 \in D(A(s))$ will be only intepreted as a limit solution to equation (1.1), but it will be a strong solution if $A(t)$ satisfies additionally the embedding property (HB) of embeddedly quasi-demi-closedness in Chapter 2 (see Section 2).

In addition to this section, there are three more sections in this chapter. Section 2 states the main results, and Section 3 obtains some preliminary estimates. Finally, Section 4 proves the main results.

The material of this chapter is based on our article [26].

2. Main Results

There are two theorems in relation to the evolution equation (1.1).

THEOREM 2.1 (Existence of a limit). *Let the nonlinear operator $A(t)$ satisfy the non-dissipativity condition (H3), the weaker range condition $(H2)'$, and the time-regulating condition (HC) or $(HC)'$. Then, for $u_0 \in D(A(s))$ and for large enough $n \in \mathbb{N}$, the quantity*

$$\prod_{i=1}^{n} J_{\frac{t}{n}}\left(s + i\frac{t}{n}\right)u_0$$

is single-valued and its limit, as $n \longrightarrow \infty$,

$$U(s+t,s)u_0 \equiv \lim_{n\to\infty} \prod_{i=1}^{n} J_{\frac{t}{n}}\left(s + i\frac{t}{n}\right)u_0$$

$$= \lim_{\mu\to 0} \prod_{i=1}^{[\frac{t}{\mu}]} J_{\mu}(s + i\mu)u_0$$

exists, if T or α_{-1} is sufficiently small, or if so are both K and α_0 where both K_D and K_E remain finite. Here $s, t \geq 0$ and $0 \leq (s+t) \leq T$, and moreover, $\alpha_{-1}, K, \alpha_0, K_D,$ and K_E are defined in Proposition 3.5 in Section 3 where how small the mentioned quantities are is also conditioned there.

This limit $U(s+t,s)u_0$ is Lipschitz continuous in $t \geq 0$ for $u_0 \in D(A(s))$.

REMARK 2.2. In loose terms, the smallness of α_{-1} means that of both M_2 (and $\tilde{M}_2$, respectively) and L_1, but the smallness of both K and α_0 means that of both M_0 (and $\tilde{M}_0$, respectively) and $\|v_0\|$, a less or equal order of which is possessed by the total variation of g on $[0,T]$, multiplied by α_2 (defined in Proposition 3.5 in Section 3). Here v_0 is any element in $A(s)u_0$, and the smallness of α_2 means that of M_4 (and $\tilde{M}_4$, respectively).

The next theorem, Theorem 2.3, concerns a limit solution and a strong solution, whose concepts are stated in Chapter 2 and will be recalled now. We make two preparations, the first of which is for a limit solution. Discretize (1.1) on $[0, T]$ as

$$u_i - \epsilon A(t_i)u_i \ni u_{i-1},$$
$$u_i \in D(A(t_i)), \tag{2.1}$$

where $n \in \mathbb{N}$ is large, and $0 < \epsilon < \lambda_0$ is such that $s \leq t_i = s + i\epsilon \leq T$ for each $i = 1, 2, \ldots, n$. Here to be noticed is that, for $u_0 \in D(A(s))$ and for small enough ϵ, u_i will exist uniquely (see Proposition 3.6 in Section 3) by hypotheses $(H2)'$, (H3), and (HC) or $(HC)'$, if the condition of smallness in Theorem 2.1 is satisfied (see also Remark 2.2).

Next, let $u_0 \in D(A(s))$, and construct the Rothe functions [**12, 32**] by defining

$$\chi^n(s) = u_0, \quad C^n(s) = A(s);$$
$$\chi^n(t) = u_i, \quad C^n(t) = A(t_i) \tag{2.2}$$
$$\text{for } t \in (t_{i-1}, t_i],$$

and

$$u^n(s) = u_0;$$
$$u^n(t) = u_{i-1} + (u_i - u_{i-1})\frac{t - t_{i-1}}{\epsilon} \tag{2.3}$$
$$\text{for } t \in (t_{i-1}, t_i] \subset [s, T].$$

Then it follows that, for some constant K_{-1},

$$\lim_{n \to \infty} \sup_{t \in [0,T]} \|u^n(t) - \chi^n(t)\| = 0;$$
$$\|u^n(t) - u^n(\tau)\| \leq K_{-1}|t - \tau|, \tag{2.4}$$

where $t, \tau \in (t_{i-1}, t_i]$, and that

$$\frac{du^n(t)}{dt} \in C^n(t)\chi^n(t);$$
$$u^n(s) = u_0, \tag{2.5}$$

where $t \in (t_{i-1}, t_i]$. Here the last equation has values in $B([s, T]; X)$, the real Banach space of all bounded functions from $[s, T]$ to X.

The second preparation is for a strong solution. Let $(Y, \|.\|_Y)$ be a real Banach space, into which the real Banach space $(X, \|.\|)$ is continuously embedded. Assume additionally that $A(t)$ satisfies the embedding condition of embeddedly quasi-demi-closedness:

(HB) If $t_n \in [0, T] \longrightarrow t$, if $x_n \in D(A(t_n)) \longrightarrow x$, and if $\|y_n\| \leq M_0$ for some $y_n \in A(t_n)x_n$ and for some positive constant M_0, then $\eta(A(t)x)$ exists and

$$|\eta(y_{n_l}) - z| \longrightarrow 0$$

for some subsequence y_{n_l} of y_n, for some $z \in \eta(A(t)x)$, and for each $\eta \in Y^* \subset X^*$, the real dual space of Y.

THEOREM 2.3 (A limit or a strong solution [**25**]). *Following Theorem 2.1, if* $u_0 \in D(A(s))$, *then the function*

$$u(t) \equiv U(t, s)u_0$$

$$= \lim_{n \to \infty} \prod_{i=1}^{n} J_{\frac{t-s}{n}}(s + i\frac{t-s}{n})u_0$$

$$= \lim_{\mu \to 0} \prod_{i=1}^{[\frac{t-s}{\mu}]} J_{\mu}(s + i\mu)u_0$$

is a limit solution of the evolution equation (1.1) on $[0, T]$, in the sense that it is also the uniform limit of $u^n(t)$ on $[0, T]$, where $u^n(t)$ satisfies (2.5).

Furthermore, if $A(t)$ satisfies the embedding property (HB), then $u(t)$ is a strong solution in Y, in the sense that

$$\frac{d}{dt}u(t) \in A(t)u(t) \quad in\ Y$$

$$for\ almost\ every\ t \in (0, T);$$

$$u(s) = u_0$$

is true. The strong solution is unique if $Y \equiv X$.

3. Some Preliminary Estimates

In order to prove the main results, Theorems 2.1 and 2.3 in Section 2, we need to prepare some preliminary estimates, which are in this section.

LEMMA 3.1. *Let* $x_i, i = 0, 1, 2, \ldots$, *be a sequence of nonnegative numbers that satisfies the difference inequality*

$$x_i \leq \lambda\gamma\alpha x_i x_{i-1} + \gamma[1 + |F(t_i) - F(t_{i-1})|]x_{i-1}, \tag{3.1}$$

$$i = 1, 2, \ldots.$$

Here

the numbers $\lambda, \omega, \gamma > 0, \alpha \geq 0$, *are such that*

$$\gamma = (1 - \lambda\omega)^{-1} > 1 \ and\ \lambda\omega < 1;$$

the function $F(t)$ *is continuous and of bounded variation on* $[0, T]$;

$0 \leq t_{i-1} < t_i \leq T$, *and* $i\lambda \leq T$.

Then either the following (P_1) *or* (P_2) *holds:*

(P_1) *There is an* $i_0 \in \{0\} \cup \mathbb{N}$, *such that* $x_{i_0} = 0$ *and such that, for all* $i \geq 0$,

$$x_i \leq \max\{x_0, x_1, \ldots, x_{i_0}\}.$$

(P_2) $x_i \neq 0$ *for all* $i \geq 0$, *and, on setting* $y_i = 1/x_i$, *we have*

$$y_i \geq e^{-\omega T}e^{-K_C}y_0 - T\alpha.$$

This y_i *is strictly greater than zero, if* x_0 *or* T *or* α *is small enough. Here* K_C *is the total variation of* F *on* $[0, T]$.

Hence, x_i *is uniformly bounded for all* $i \geq 0$, *if* T *or* α *or* x_0 *is small enough.*

PROOF. We divide the proof into two cases.

Case 1, where $x_i \neq 0$ **for all** i. On dividing (3.1) by $x_i x_{i-1}$ and setting $y_i = 1/x_i$, we have

$$y_i \geq c_i y_{i-1} - d_i,$$

where

$$c_i = \gamma^{-1}e^{-|F(t_i) - F(t_{i-1})|} \leq 1;$$

$$d_i = \lambda \alpha e^{-|F(t_i) - F(t_{i-1})|} \leq \lambda \alpha.$$

Here we have used $[1 + |F(t_i) - F(t_{i-1})|] \leq e^{|F(t_i) - F(t_{i-1})|}$.

Solving this linear difference inequality yields

$$
\begin{aligned}
y_i &\geq (c_i c_{i-1} \cdots c_1) y_0 - [(c_i c_{i-1} \cdots c_2) d_1 + \cdots + c_i d_{i-1} + d_i] \\
&\geq \gamma^{-i} e^{-K_C} y_0 - (i\lambda) \alpha \\
&\geq e^{-\omega T} e^{-K_C} y_0 - T\alpha,
\end{aligned}
\tag{3.2}
$$

which is strictly greater than zero, if x_0 or T or α is sufficiently small. Here $i\lambda \leq T$, and K_C is the total variation of $F(t)$ on $[0, T]$.

Case 2, where $x_{i_0} = 0$ for some $i_0 \in \{0\} \cup \mathbb{N}$. Then clearly, (3.1) implies $x_i = 0$ for $i \geq i_0$, so for all $i \geq 0$,

$$x_i \leq \max\{x_0, x_1, \ldots, x_{i_0}\}.$$

The proof is complete. □

LEMMA 3.2. *Let α_0 be a positive number, and let the x_i in Lemma 3.1 satisfy instead the difference inequality*

$$
\begin{aligned}
x_i \leq{} &\lambda \gamma \alpha x_i x_{i-1} + \gamma [1 + |F(t_i) - F(t_{i-1})|] x_{i-1} \\
&+ \gamma |G(t_i) - G(t_{i-1})|, \quad i = 1, 2, \ldots.
\end{aligned}
\tag{3.3}
$$

Here $\lambda, \omega, \alpha, \gamma$ and $F(t)$ are as in Lemma 3.1, and the function $G(t)$ is continuous and of bounded variation on $[0, T]$. Then the following (P_3) holds:

(P_3) *If $x_0 \leq \alpha_0$, then, for all $i \geq 1$,*

$$x_i \leq e^{\omega T} \alpha_0 e^{K_D} / [1 - T\alpha \alpha_0 e^{\omega T} e^{K_D}],$$

if T or α is small enough. This is also true for small enough α_0, where K_D should be finite however small α_0 is. Here $i\lambda \leq T$, and K_D is the sum of the total variation of $F(t)$ and that of $G(t)/\alpha_0$, respectively, on $[0, T]$. Thus, a condition for K_D to be finite, regardless of the small value of α_0, is to require that the total variation K_G of $G(t)$ on $[0, T]$ be small enough, so that K_G/α_0 is finite. For instance, assume that K_G is proportional to α_0.

PROOF. Let $x_0 \leq \alpha_0$. Then it follows from (3.3) that, for $|h(t) - h(\tau)| \equiv |F(t) - F(\tau)| + |G(t) - G(\tau)|/\alpha_0$,

$$
\begin{aligned}
x_1 &\leq \frac{\gamma \alpha_0 [1 + |h(t_1) - h(t_0)|]}{1 - \lambda \alpha \alpha_0 \gamma} \\
&\leq \frac{\gamma \alpha_0 e^{|h(t_1) - h(t_0)|}}{1 - \lambda \alpha \alpha_0 \gamma},
\end{aligned}
$$

if $\lambda \alpha \alpha_0$ is small enough. Since the right hand side R_d of the above is greater than or equal to α_0, we have

$$\gamma |G(t_2) - G(t_1)| \leq \gamma [|G(t_2) - G(t_1)|/\alpha_0] R_d.$$

Hence (3.3) delivers

$$
\begin{aligned}
x_2 &\leq \frac{\gamma^2 \alpha_0 [1 + |h(t_2) - h(t_1)|][1 + |h(t_1) - h(t_0)|]}{1 - \lambda \alpha \alpha_0 \gamma - \lambda \alpha \alpha_0 \gamma^2 [1 + |h(t_1) - h(t_0)|]} \\
&\leq \frac{\gamma^2 \alpha_0 e^{[|h(t_2) - h(t_1)| + |h(t_1) - h(t_0)|]}}{1 - \lambda \alpha \alpha_0 \gamma - \lambda \alpha \alpha_0 \gamma^2 e^{|h(t_1) - h(t_0)|}},
\end{aligned}
$$

if $\lambda\alpha\alpha_0$ is small enough. Again, the right hand side of the above is greater than or equal to α_0, so, as in estimating x_2, we derive, on using $[1 + |h(t) - h(\tau)|] \leq e^{|h(t)-h(\tau)|}$,

$$x_3 \leq \frac{\gamma^3 \alpha_0 e^{[|h(t_3)-h(t_2)|+|h(t_2)-h(t_1)|+|h(t_1)-h(t_0)|]}}{1 - \lambda\alpha\alpha_0\gamma\{1 + \gamma e^{|h(t_1)-h(t_0)|} + \gamma^2 e^{[|h(t_2)-h(t_1)|+|h(t_1)-h(t_0)|]}\}},$$

if $\lambda\alpha\alpha_0$ is small enough. Clearly, the right side of the above is greater than or equal to α_0.

Continued in this way, the following is obtained

$$\begin{aligned}
x_i &\leq \frac{\gamma^i \alpha_0 e^{\sum_{j=0}^{i-1} |h(t_{j+1})-h(t_j)|}}{1 - \lambda\alpha\alpha_0\gamma[1 + \sum_{j=1}^{i-1} \gamma^j e^{\sum_{k=0}^{j-1} |h(t_{k+1})-h(t_k)|}]} \\
&\leq \gamma^i \alpha_0 e^{K_D}/[1 - \lambda\alpha\alpha_0(i\gamma^i e^{K_D})] \\
&\leq e^{\omega T}\alpha_0 e^{K_D}/[1 - T\alpha\alpha_0 e^{\omega T} e^{K_D}],
\end{aligned}$$

(3.4)

if small enough is T or α, or if so is α_0 where K_D remains finite. Here $i\lambda \leq T$, and K_D is the sum of the total variation of $F(t)$ and that of $G(t)/\alpha_0$, respectively, on $[0, T]$.

The proof is complete.   □

LEMMA 3.3. *Let $u_0 \in D(A(s))$, and let the weaker range condition $(H2)'$ hold. Let $0 < \lambda < \lambda_0$ and $0 \leq t_i = s+i\lambda \leq T$. Then there is a sequence $u_i \in D(A(t_i)), i = 1, 2, \ldots$, such that*

$$u_i - \lambda A(t_i)u_i \ni u_{i-1}.$$

PROOF. By $(H2)'$, there is a $u_1 \in D(A(t_1))$ that satisfies

$$u_1 - \lambda A(t_1)u_1 \ni u_0.$$

For this u_1, we have, by $(H2)'$ again, a $u_2 \in D(A(t_2))$ that satisfies

$$u_2 - \lambda A(t_2)u_2 \ni u_1.$$

Continuing in this way, there are $u_i \in D(A(t_i)), i = 1, 2, \ldots$, such that

$$u_i - \lambda A(t_i)u_i \ni u_{i-1}.$$

This completes the proof.   □

LEMMA 3.4. *Let the weaker range condition $(H2)'$ and the non-dissipativity condition $(H3)$ hold. Then the u_i in Lemma 3.3 satisfies, for all $i \geq 0$ and $\eta = 1/(1 - \lambda\delta_0)$,*

$$\|u_i\| \leq \begin{cases} \eta^i \tilde{a} + \eta^i \|u_0\| + (i\lambda)\eta^i a, & \text{if } \delta_0 > 0; \\ \tilde{a} + \|u_0\| + (i\lambda)a, & \text{if } \delta_0 \leq 0. \end{cases}$$

(3.5)

Hence, $\|u_i\|$ is uniformly bounded for all i, λ.

PROOF. The proof will be made with the case $\delta_0 > 0$ for which $\eta > 1$, as the othe case is similar.

The method of induction will complete the proof, as the arguments below show. If $i = 1$, we have, by the nondissipativity condition $(H1)'$,

$$\|u_1\| \leq (\eta\|u_0\| + \lambda\eta a) \text{ or } \tilde{a},$$

which is $\leq$ the right side of (3.5) for $i = 1$. Now we show (3.5) is true for $i = i$ when it is so for $i = i - 1$. By the non-dissipativity condition $(H1)'$, we have

$$\|u_i\| \leq (\eta\|u_{i-1}\| + \eta\lambda a) \text{ or } \tilde{a},$$

which, combined with the induction assumption, concludes the proof. Here used was $1 \leq \eta \leq \eta^i$ for $\eta > 1$. □

In view of (3.4), we are led to the claim:

PROPOSITION 3.5. *Let the weaker range condition* $(H2)'$, *the non-dissipativity condition (H3), and the time-regulating condtion (HC) or* $(HC)'$ *hold. Then, under (HC), the* u_i *in Lemma 3.3 satisfies, for small enough* $\lambda > 0$, *for* $i = 0, 1, 2, \ldots$, *and for* $u_{-1} = u_0 - \lambda v_0$, *where* v_0 *is any element in* $A(s)u_0$, *the following (Q1), (Q2), (Q3), and (Q4):*

(Q1) There is a positive number ω_0, *such that*

$$L_0(\|u_i\|, \|u_{i-1}\|), \quad M_1(\|u_i\|, \|u_{i-1}\|) \leq \omega_0.$$

(Q2) There are one positive number $K > 0$ *and three nonnegative numbers* $\alpha_{-1}, \alpha_1, \alpha_2 \geq 0$, *such that*

$$M_0(\|u_i\|, \|u_{i-1}\|) \leq K; \quad M_2(\|u_i\|, \|u_{i-1}\|) \leq \alpha_{-1};$$
$$L_1(\|u_i\|, \|u_{i-1}\|) \leq \alpha_{-1};$$
$$M_3(\|u_i\|, \|u_{i-1}\|) \leq \alpha_1; \quad M_4(\|u_i\|, \|u_{i-1}\|) \leq \alpha_2.$$

(Q3) Assume $\|v_0\| \leq \alpha_0$ *for some number* $\alpha_0 > 0$. *Then the inequality is true*

$$\|u_i - u_{i-1}\|/\lambda \leq \quad [K \text{ or } K_1(i) \text{ or } K_2(i) \text{ or } \cdots \text{ or } K_{i-1}(i) \text{ or } \tilde{K}_i],$$

where the right side is interpreted as K *or* $\tilde{K}_0$ *when* $i = 0$, *and as* K *or* $\tilde{K}_1$ *when* $i = 1$. *Here, for* $j = 1, 2, \ldots, i - 1$, *and for small* λ *so that* $\gamma = (1 - \lambda\omega_0)^{-1}$ *exists,*

$$K_j(i) = \frac{\gamma^j K e^{\sum_{k=i-j}^{i-1} |h(t_{k+1})-h(t_k)|}}{1 - \lambda\alpha_{-1}K\gamma[1 + \sum_{k=1}^{j-1} \gamma^k e^{\sum_{l=i-k-1}^{i-2} |h(t_{l+1})-h(t_l)|}]}$$
$$\leq \gamma^i K e^{K_D}/[1 - \lambda\alpha_{-1}K(i\gamma^i e^{K_D})]$$
$$\leq e^{\omega_0 T} K e^{K_D}/[1 - T\alpha_{-1} K e^{\omega_0 T} e^{K_D}],$$

provided that T *or* α_{-1} *is small enough, or that* K *is so where* K_D *remains finite. Here* $i\lambda \leq T - s$, $|h(t) - h(\tau)| = |f(t) - f(\tau)|\alpha_1 + |g(t) - g(\tau)|\alpha_2/K$, *and* K_D *is the sum of the total variation of* $f(t)\alpha_1$ *and that of* $g(t)\alpha_2/K$, *respectively, on* $[0, T]$. *Furthermore,* $\tilde{K}_0 = \alpha_0$, *and*

$$\tilde{K}_i = \frac{\gamma^i \alpha_0 e^{\sum_{j=0}^{i-1} |H(t_{j+1})-H(t_j)|}}{1 - \lambda\alpha_{-1}\alpha_0\gamma[1 + \sum_{j=1}^{i-1} \gamma^j e^{\sum_{k=0}^{j-1} |H(t_{k+1})-H(t_k)|}]}$$
$$\leq \gamma^i \alpha_0 e^{K_E}/[1 - \lambda\alpha_{-1}\alpha_0(i\gamma^i e^{K_E})]$$
$$\leq e^{\omega_0 T} \alpha_0 e^{K_E}/[1 - T\alpha_{-1}\alpha_0 e^{\omega_0 T} e^{K_E}],$$

if T *or* α_{-1} *is small enough, or if so is* α_0 *where* K_E *remains finite. Here* $i\lambda \leq T-s$, $|H(t) - H(\tau)| = |f(t) - f(\tau)|\alpha_1 + |g(t) - g(\tau)|\alpha_2/\alpha_0$, *and* K_E *is the total variation of* $f(t)\alpha_1$ *and that of* $g(t)\alpha_2/\alpha_0$, *respectively, on* $[0, T]$.

Thus, a condition for both K_D *and* K_E *to be finite, regardless of the small value of* K *and* α_0, *is to require that either* α_2 *or the total variation* K_G *of* $g(t)$ *on*

$[0, T]$ be small enough, so that both $K_G \alpha_2/K$ and $K_G \alpha_2/\alpha_0$ are finite. For example, assume that either α_2 or K_G is proportional to $K\alpha_0$.

In the above $K_j(i)$ and $\tilde{K}_i$, the convention was used that $\sum_{l=k}^m a_l = 0$ for $m < k$.

Hence, there is a constant $N_1 > K, \alpha_0$, such that

$$\|u_i - u_{i-1}\|/\lambda \le N_1, \tag{3.6}$$

uniformly bounded for all i, λ, if T or α_{-1} is sufficiently small, or if so are both K and α_0 where both K_D and K_E remain finite. Here the constant N_1 depends on $K, \omega_0, T, \alpha_{-1}, \alpha_0, K_D$, and K_E.

(Q4)

$$\|u_i - u_0\| \le \eta^i K(i\lambda) + [\eta^{i-1}b_1 + \eta^{i-2}b_2 + \cdots + \eta b_{i-1} + b_i],$$

where $\eta = 1/[1 - \lambda(\omega_0 + \alpha_{-1}\|v_0\|)]$ exists for small λ, and

$$b_i = \eta\lambda\|v_0\| + \eta\lambda[|f(t_i) - f(t_0)|\alpha_1\|v_0\| + |g(t_i) - g(t_0)|\alpha_2].$$

Hence, there is a constant N_2 such that

$$\|u_i - u_0\| \le N_2[1 - \lambda(\omega_0 + \alpha_{-1}\|v_0\|)]^{-i}(2i + 1)\lambda$$
$$\le N_2[1 - \lambda(\omega_0 + \alpha_{-1}N_1)]^{-i}(2i + 1)\lambda \tag{3.7}$$
$$\text{because } \|v_0\| \le \alpha_0 < N_1,$$

where the constant N_2 depends on $K, \|v_0\|$, and the sum K_F of the total variation of $f(t)\alpha_1\|v_0\|$ and that of $g(t)\alpha_2$, respectively, on $[0, T]$.

On the other hand, the same results are obtained under $(HC)'$, if the above $M_k, k = 0, \ldots, 4$, are replaced by $\tilde{M}_k$'s.

PROOF. The proof will be made with (HC) assumed, as it is similar under $(HC)'$.

We divide the proof into four steps, where the simple inequality, for $c > 0$, $(1 + c) \le e^c$ will be frequently utilized.

Step 1. The assertions (Q1) and (Q2) follow immediately from Lemma 3.4.

Step 2. To prove (Q3), the method of mathematical induction will be used. Since

$$u_1 - \lambda A(t_1)u_1 \ni u_0; \quad u_0 - \lambda v_0 \equiv u_{-1},$$

(Q3) is true for $i = 1$ by the time-regulating condition (HC). This is because, for $x_1 \equiv \|u_1 - u_0\|/\lambda$, we have

$$x_1 \le K \quad \text{if } (u_1 - u_0) \in S_1(\lambda);$$

but, if $(u_1 - u_0) \in S_2(\lambda)$, then

$$x_1 \le \lambda\gamma\alpha_{-1}\alpha_0 x_1 + \gamma[1 + |f(t_1) - f(t_0)|\alpha_1]\alpha_0$$
$$+ \gamma|g(t_1) - g(t_0)|(\alpha_2/\alpha_0)\alpha_0,$$

so $x_1 \le \tilde{K}_1$. Here $\gamma \equiv 1/(1 - \lambda\omega_0)$ for small λ.

Step 3. Assume that (Q3) is true for $i = i - 1$, and we prove it is also true for $i = i$. This will follow from (HC), combined with the induction assumption, as the following arguments show. If $(u_i - u_{i-1}) \in S_1(\lambda)$, then

$$\|u_i - u_{i-1}\| \le K(t_i - t_{i-1}) = K\lambda.$$

But if $(u_i - u_{i-1}) \in S_2(\lambda)$, then, for $x_i \equiv \|u_i - u_{i-1}\|/\lambda$,

$$x_i \leq \lambda\gamma\alpha_{-1}x_ix_{i-1} + \gamma[1 + |f(t_i) - f(t_{i-1})|\alpha_1]x_{i-1}$$
$$+ \gamma|g(t_i) - g(t_{i-1})|\alpha_2,$$

which, in conjuction with the induction assumption, yields (Q3) for $i = i$. This is because

$$x_{i-1} \leq [K \text{ or } K_1(i-1) \text{ or } K_2(i-1) \text{ or } \cdots \text{ or } K_{i-1}(i-1) \text{ or } \tilde{K}_{i-1}];$$
$$0 < K \leq [K \text{ and } K_j(i-1), j = 1, 2, \ldots, i-1];$$
$$\alpha_0 \leq \tilde{K}_{i-1},$$

so that

$$\gamma|g(t_i) - g(t_{i-1})|\alpha_2 \leq \gamma|g(t_i) - g(t_{i-1})|(\alpha_2/K)$$
$$[K \text{ and } K_j(i-1), j = 1, 2, \ldots, i-1];$$
$$\gamma|g(t_i) - g(t_{i-1})|\alpha_2 \leq \gamma|g(t_i) - g(t_{i-1})|(\alpha_2/\alpha_0)\tilde{K}_{i-1}.$$

Step 4. Again, (Q4) will be proved by induction. (Q4) is correct by (HC) if $i = 1$, because, for $u_{-1} = u_0 - \lambda v_0$,

$$\|u_1 - u_0\| \leq K\lambda \leq \eta K\lambda \quad \text{if } (u_1 - u_0) \in S_1(\lambda);$$
$$\|u_1 - u_0\| \leq b_1 \quad \text{if } (u_1 - u_0) \in S_2(\lambda).$$

Next, by assuming that (Q4) is correct for $i = i - 1$, we shall show that it is also correct for $i = i$. This follows from (HC), together with induction assumption. For

$$\|u_i - u_0\| \leq K(i\lambda) \leq \eta^i K(i\lambda) \quad \text{if } (u_i - u_0) \in S_1(\lambda);$$

but, if $(u_i - u_0) \in S_2(\lambda)$, then we have, for the η and b_i in (Q4),

$$\|u_i - u_0\| \leq \eta\|u_{i-1} - u_0\| + b_i$$
$$\leq \eta\{\eta^{i-1}K[(i-1)\lambda]$$
$$+ [\eta^{i-2}b_1 + \eta^{i-3}b_2 + \cdots + \eta b_{i-2} + b_{i-1}]\} + b_i,$$

which is less than or equal to the right side of (Q4) with $i = i$.

The proof is complete.   □

PROPOSITION 3.6. *Let $A(t)$ satisfy the weaker range condition (H2)′, the non-dissipativity condition (H3), and the time-regulating condition (HC) or (HC)′. Then the u_i in Lemma 3.3 is unique, if, as is conditioned in Proposition 3.5, T or α_{-1} is small enough, or if so are both K and α_0 where both K_D and K_E remain finite. Thus $u_i = J_\lambda(t_i)u_0$.*

PROOF. We will show that $u_i = v_i$, if $v_i \in D(A(t_i)), i = 1, 2, \ldots$, is another sequence such that

$$v_i - \lambda A(t_i)v_i \ni v_{i-1}, \quad i = 1, 2, \ldots,$$
$$v_0 = u_0.$$

But this follows from the non-dissipativity condition (H3) because, then,

$$\|u_i - v_i\| \leq [\|u_{i-1} - v_{i-1}\| + \lambda L_0(\|u_i\|, \|v_i\|)\|u_i - v_i\|$$
$$+ \lambda L_1(\|u_i\|, \|v_i\|)(\|v_i - v_{i-1}\|/\lambda)\|u_i - v_i\|],$$

whence, for some constant ξ with $\lambda\xi$ small enough,

$$\|u_i - v_i\| \leq [1/(1 - \lambda\xi)]\|u_{i-1} - v_{i-1}\|.$$

This inequality concludes the proof, where $\|u_0 - v_0\| = 0$. Here used was bounded-ness of $\|u_i\|, \|v_i\|$, and $\|v_i - v_{i-1}\|/\lambda$ under smallness of T or α_{-1} or both K and α_0, by Lemma 3.4 and Proposition 3.5. $\qquad\Box$

It follows immediately from the proof of Propositions 3.5 that

COROLLARY 3.7. *Let $u_0 \in D(A(s))$ and let $u_i, i = 1, 2, \ldots$, satisfy the difference relation, where $0 < \lambda < \lambda_0$,*

$$u_i - \lambda A(t_i)u_i \ni u_{i-1}, \quad i = 1, 2, \ldots.$$

Then the (Q3) and (Q4) in Propositions 3.5 are, respectively, still true, if we do not assume that $A(t)$ satisfies the non-dissipativity condition (H3), the weaker range condition (H2)', and the time-regulating condition (HC) or (HC)', but assume that u_i satisfies, respectively, both the conditions (D1) and (D2), and both the conditions (D3) and (D4):

(D1) *For $x_1 \equiv \|u_1 - u_0\|/\lambda$,*

$$x_1 \leq \begin{cases} K, \quad or \\ \lambda\gamma\alpha_{-1}\alpha_0 x_1 + \gamma[1 + |f(t_1) - f(t_0)|\alpha_1]\alpha_0 \\ \quad + \gamma|g(t_1) - g(t_0)|(\alpha_2/\alpha_0)\alpha_0. \end{cases}$$

(D2) *For $x_i \equiv \|u_i - u_{i-1}\|/\lambda$,*

$$x_i \leq \begin{cases} K, \quad or \\ \lambda\gamma\alpha_{-1}x_i x_{i-1} + \gamma[1 + |f(t_i) - f(t_{i-1})|\alpha_1]x_{i-1} \\ \quad + \gamma|g(t_i) - g(t_{i-1})|\alpha_2. \end{cases}$$

(D3)

$$\|u_1 - u_0\| \leq \begin{cases} K\lambda, \quad or \\ b_1. \end{cases}$$

(D4)

$$\|u_i - u_0\| \leq \begin{cases} K(i\lambda), \quad or \\ \eta\|u_{i-1} - u_0\| + b_i. \end{cases}$$

Here

v_0 *is any element in $A(s)u_0$;*

$u_0 - \epsilon v_0 = u_{-1}$;

$\eta = 1/[1 - \lambda(\omega_0 + \alpha_{-1}\|v_0\|)]$;

$b_i = \eta\lambda\|v_0\| + \eta\lambda[|f(t_i) - f(t_0)|\alpha_1\|v_0\| + |g(t_i) - g(t_0)|\alpha_2]$;

the functions f and g are as in (HC) or (HC)'.

4. Proof of the Main Results

Proof of Theorem 2.1 will be done after those of Propositions 4.1 and 4.2 below, where the formulation of these two propositions resemble that of Propositions 5.1 and 5.4 in Chapter 2.

We are enabled to obtain Proposition 4.1 below, as in the proof of Proposition 5.1 in Section 6 of Chapter 2, by the preliminary estimates in Section 3, together with the difference equation theory in Chapter 2.

PROPOSITION 4.1. *Under the assumptions of Proposition 4.2, the inequality is true*

$$a_{m,n} \leq \begin{cases} K|n\mu - m\lambda|, & \text{if } S_2(\mu) = \emptyset; \\ c_{m,n} + s_{m,n} + d_{m,n} + f_{m,n} + g_{m,n}, & \text{if } S_1(\mu) = \emptyset; \end{cases}$$

where $a_{m,n}, c_{m,n}, s_{m,n}, f_{m,n}, g_{m,n}$ and K are defined in Proposition 4.2.

In view of this and Proposition 3.5, we are led to the claim:

PROPOSITION 4.2. *Let $x \in D(A(s))$ where $0 \leq s \leq T$, and let small enough $\lambda, \mu > 0$ and let $n, m \in \mathbb{N}$, be such that $0 \leq (s + m\lambda), (s + n\mu) \leq T$, and such that $\lambda_0 > \lambda \geq \mu > 0$. Let $A(t)$ satisfy the non-dissipativity condition (H3), the weaker range condition (H2)', and the time-regulating condition (HC) or (HC)'. Then the inequality is true if, as is conditioned in Proposition 3.5, T or α_{-1} is sufficiently small, or if so are both K and α_0 where K_D and K_E remain finite:*

$$a_{m,n} \leq c_{m,n} + s_{m,n} + d_{m,n} + e_{m,n} + f_{m,n} + g_{m,n}. \tag{4.1}$$

Here

$\alpha_{-1}, K,$ and α_0 are defined in Proposition 3.5;

$$a_{m,n} \equiv \| \prod_{i=1}^{n} J_\mu(s + i\mu)x - \prod_{i=1}^{m} J_\lambda(s + i\lambda)x \|$$

exists by Proposition 3.6;

$$\gamma \equiv [1 - \mu(\omega_0 + \alpha_{-1}N_1]^{-1} > 1; \quad \alpha \equiv \frac{\mu}{\lambda}; \quad \beta \equiv 1 - \alpha;$$

$$c_{m,n} = 2N_2\gamma^n[(n\mu - m\lambda) + \sqrt{(n\mu - m\lambda)^2 + (n\mu)(\lambda - \mu)}];$$

$$s_{m,n} = 2N_2\gamma^n(1 - \lambda\omega)^{-m}\sqrt{(n\mu - m\lambda)^2 + (n\mu)(\lambda - \mu)};$$

$$d_{m,n} = [N_3\rho(\delta)\gamma^n(m\lambda)] + \{N_3\frac{\rho(T)}{\delta^2}\gamma^n[(m\lambda)(n\mu - m\lambda)^2$$

$$+ (\lambda - \mu)\frac{m(m+1)}{2}\lambda^2]\};$$

$$e_{m,n} = K\gamma^n\sqrt{(n\mu - m\lambda)^2 + (n\mu)(\lambda - \mu)};$$

$$f_{m,n} = N_2[\gamma^n\mu + \gamma^n(1 - \lambda\omega)^{-m}\lambda];$$

$$g_{m,n} = N_3\rho(|\lambda - \mu|)\gamma^n(m\lambda);$$

$$N_3 = \gamma(\alpha_1 N_1 + \alpha_2); \quad \delta > 0 \quad \text{is arbitrary;}$$

$$\rho(r) \equiv \rho_1(r) + \rho_2(r);$$

$$\rho_1(r) \equiv \sup\{|f(t) - f(\tau)| : 0 \leq t, \tau \leq T, |t - \tau| \leq r\};$$

$$\rho_2(r) \equiv \sup\{|g(t) - g(\tau)| : 0 \leq t, \tau \leq T, |t - \tau| \leq r\};$$

ρ_1 and ρ_2 are the modulus of continuity of

f and g, respectively, on $[0, T]$;

$\omega_0, \alpha_{-1}, N_1$, and N_2 are defined in Proposition 3.5.

PROOF. We will use the method of mathematical induction and divide the proof into two steps.

Step 1. (4.1) is clearly true by (3.7) in Proposition 3.5, if $(m, n) = (0, n)$ or $(m, n) = (m, 0)$.

Step 2. By assuming that (4.1) is true for $(m, n) = (m-1, n-1)$ or $(m, n) = (m, n-1)$, we will show that it is also true for $(m, n) = (m, n)$. This is done by the arguments below.

Using the nonlinear resolvent identity in Lemma 4.3 in Chapter 1, we have

$$a_{m,n} = \| J_u(s + n\mu) \prod_{i=1}^{n-1} J_\mu(s + i\mu)x - J_\mu(s + m\lambda)$$

$$[\alpha \prod_{i=1}^{m-1} J_\lambda(s + i\lambda)x + \beta \prod_{i=1}^{m} J_\lambda(s + i\lambda)x)]\|.$$

Here $\alpha = \frac{\mu}{\lambda}$ and $\beta = \frac{\lambda - \mu}{\lambda}$.

Under the time-regulating condition (HC) or $(HC)'$, it follows that, if the element inside the norm of the right side of the above equality is in $S_1(\mu)$, then, by (Q2) in Proposition 3.5,

$$a_{m,n} \le K|m\lambda - n\mu|,$$

which is less than or equal to the right side of (4.1) with $(m, n) = (m, n)$, where

$$\gamma^n = [1 - \mu(\omega_0 + \alpha_{-1}N_1)]^{-n} > 1.$$

If that element instead lies in $S_2(\mu)$, then, by (Q1), (Q2), and (3.6) in Proposition 3.5,

$$a_{m,n} \le \gamma(\alpha a_{m-1,n-1} + \beta a_{m,n-1}) +$$
$$\gamma\mu[|f(s + m\lambda) - f(s + n\mu)|\alpha_1 N_1$$
$$+ |g(s + m\lambda) - g(s + n\mu)|\alpha_2]$$
$$\le [\gamma\alpha a_{m-1,n-1} + \gamma\beta a_{m,n-1}] + N_3\mu\rho(|n\mu - m\lambda|),$$

where $N_3 = \gamma(\alpha_1 N_1 + \alpha_2)$, $\rho(r) = \rho_1(r) + \rho_2(r)$, and $\rho_1(r)$ and $\rho_2(r)$ are the modulus of continuity of f and g, respectively, on $[0, T]$. From here on, the rest of the proof can be completed by being patterned after that of Proposition 5.2 in Chapter 2.   □

We are ready for **Proof of Theorem 2.1:**

PROOF. Let the quantities T or α_{-1} or both K and α_0 in Proposition 3.5 be small where K_D and K_E remain finite. Then, for $x \in D(A(s))$, it follows from Proposition 4.2, by setting $\mu = \frac{t}{n} \le \lambda = \frac{t}{m} < \lambda_0$, and $\delta^2 = \sqrt{\lambda - \mu}$, that, as $n, m \longrightarrow \infty$, $a_{m,n}$ converges to 0 uniformly for $0 \le (s + t) \le T$. Thus

$$\lim_{n \to \infty} \prod_{i=1}^{n} J_{\frac{t}{n}}(s + i\frac{t}{n})x$$

exists for $x \in D(A(s))$.

On the other hand, setting $\mu = \lambda = \frac{t}{n} < \lambda_0, m = [\frac{t}{\mu}]$ and setting $\delta^2 = \sqrt{\lambda - \mu}$, it follows that

$$\lim_{n \to \infty} \prod_{i=1}^{n} J_{\frac{t}{n}}(s + i\frac{t}{n})u_0 = \lim_{\mu \to 0} \prod_{i=1}^{[\frac{t}{\mu}]} J_\mu(s + i\mu)u_0. \tag{4.2}$$

Now, to show the Lipschitz property, (4.2) and Crandall-Pazy [**8**, Page 71] will be used. From (3.6) in Proposition 3.5, it is derived that

$$\|u_n - u_m\|$$
$$\leq \|u_n - u_{n-1}\| + \|u_{n-1} - u_{n-2}\| + \ldots + \|u_{m+1} - u_m\|$$
$$\leq N_1\mu(n - m) \quad \text{for } x \in D(A(s));$$

$$u_n = \prod_{i=1}^{n} J_\mu(s + i\mu)x; \quad u_m = \prod_{i=1}^{m} J_\mu(s + i\mu)x.$$

Here $n = [\frac{t}{\mu}], m = [\frac{\tau}{\mu}], t > \tau$ and $0 < \mu < \lambda_0$. The proof is completed by making $\mu \longrightarrow 0$ and using (4.2). $\qquad\square$

Proof of Theorem 2.3 will be done after those of Propositions 4.3 and 4.4 below. Use will be made of the setup in (2.1), (2.2), and (2.3). Due to

$$\|u_i - u_{i-1}\|/\epsilon \leq N_1$$

by Proposition 3.5, we have the results in Propositions 4.3 and 4.4 below, as in the proof of Propositions 5.3 and 5.4 in Chapter 2.

PROPOSITION 4.3. *For $u_0 \in D(A(s))$, we have that*

$$\lim_{n \to \infty} \sup_{t \in [0, T]} \|u^n(t) - \chi^n(t)\| = 0;$$

$$\|u^n(t) - u^n(\tau)\| \leq N_1|t - \tau|,$$

where $t, \tau \in (t_{i-1}, t_i]$, and that

$$\frac{du^n(t)}{dt} \in C^n(t)\chi^n(t);$$
$$u^n(s) = u_0,$$

where $t \in (t_{i-1}, t_i]$. Here the last equation has values in $B([s, T]; X)$, the real Banach space of all bounded functions from $[s, T]$ to X.

PROPOSITION 4.4. *If $A(t)$ satisfies the assumptions in Theorem 2.1, then*

$$\lim_{n \to \infty} u^n(t) = \lim_{n \to \infty} \prod_{i=1}^{n} J_{\frac{t-s}{n}}(s + i\frac{t-s}{n})u_0$$

$$= \lim_{\mu \to 0} \prod_{i=1}^{[\frac{t-s}{\mu}]} J_\mu(s + i\mu)u_0$$

uniformly for finite $0 \leq (s + t) \leq T$ and for $u_0 \in D(A(s))$.

Here is **Proof of Theorem 2.3:**

PROOF. That $u(t)$ is a limit solution follows from Propositions 4.3 and 4.4. That $u(t)$ is a strong solution under the embedding property (HB) follows as in the Step 5 for the proof of Theorem 2.5 in Chapter 1. $\qquad\square$

CHAPTER 9

Nonlinear Non-autonomous Parabolic Equations (II)

1. Introduction

In this chapter, nonlinear non-autonomous, parabolic initial-boundary value problems with the u dependence in the coefficient function of the leading term will be solved by using the results in Chapter 8. The obtained solutions will be strong ones under suitable assumptions.

Thus, by allowing the u dependence in the coefficient function $\alpha(x, u, u_x)$, we shall extend the problem in (1.1) of Chapter 6 to this nonlinear, non-autonomous equation with the nonlinear Robin boundary condition, as well as to its higher space dimensional analogue

$$
\begin{aligned}
&u_t(x, t) = \alpha(x, t, u, u_x) u_{xx}(x, t) + \tilde{g}(x, t, u, u_x), \\
&\qquad\qquad (x, t) \in (0, 1) \times (0, T); \\
&u_x(0, t) \in \beta_0(u(0, t)), \quad u_x(1, t) \in -\beta_1(u(1, t)); \\
&u(x, 0) = u_0(x).
\end{aligned}
\tag{1.1}
$$

Here, as before, made are the similar assumptions:

- $\beta_0, \beta_1 : \mathbb{R} \longrightarrow \mathbb{R}$, are multi-valued, maximal monotone functions with $0 \in \beta_0(0) \cap \beta_1(0)$.
- $\alpha(x, t, z, p)$ and $\tilde{g}(x, t, z, p)$ are real-valued, continuous functions of their arguments $x \in [0, 1], t \in [0, T], z \in \mathbb{R}$, and $p \in \mathbb{R}$. Here $T > 0$.
- $\alpha(x, t, z, p)$ is greater than or equal to some positive constant δ_0 for all its arguments x, t, and p.
- $\tilde{g}(x, t, z, p)$ satisfies, for some real constant δ_{00} and some positive constant $a > 0$,

$$
z\tilde{g}(x, t, z, p) \leq \delta_{00}|z|^2 + |z|a.
$$

- $\tilde{g}(x, t, z, p)$ is of at most linear growth in p, that is, for some nonnegative, continuous function $M_{00}(x, t, z)$,

$$
|\tilde{g}(x, t, z, p)| \leq M_{00}(x, t, z)(1 + |p|).
$$

- The following are true for some continuous, nonnegative functions N_{00} and N_{01} and for some continuous function ζ of bounded variation:

$$
\begin{aligned}
|\alpha(x, t, z, p) &- \alpha(x, \tau, z, p)|/\alpha(x, \tau, z, p) \\
&\leq |\zeta(t) - \zeta(\tau)| N_{00}(x, t, \tau, z); \\
|\tilde{g}(x, t, z, p) - \tilde{g}(x, \tau, z, p)| &\leq |\zeta(t) - \zeta(\tau)| N_{01}(x, t, \tau, z)(1 + |p|).
\end{aligned}
$$

- The following are true for some continuous, nonnegative functions N_{02} and N_{03}:

$$|\alpha(x,t,z_1,p) - \alpha(x,t,z_2,p)|/\alpha(x,\tau,z_2,p)$$
$$\leq N_{02}(x,t,\tau,z_1,z_2)|z_1 - z_2|;$$
$$|\tilde{g}(x,t,z_1,p) - \tilde{g}(x,t,z_2,p)| \leq N_{03}(x,t,z_1,z_2)(1+|p|)|z_1 - z_2|.$$

To be observed is that (1.1) is different from (1.1) of Chapter 6, in that here $\alpha(x,t,z,p)$ has the z dependence, and $\tilde{g}(x,t,z,p)$ is not necessarily monotone non-increasing in z.

(1.1) will be written as the nonlinear evoution

$$\frac{d}{dt}u(t) = H(t)u(t), \quad t \in (0,T);$$
$$u(0) = u_0,$$

(1.2)

where the nonlinear, multi-valued operator

$$H(t) : D(H(t)) \subset C[0,1] \longrightarrow C[0,1]$$

is defined by

$$H(t)v = \alpha(x,t,v,v')v'' + \tilde{g}(x,t,v,v');$$
$$v \in D(H(t)) \equiv \{w \in C^2[0,1] : w'(j) \in (-1)^j \beta_j w(j), j = 0,1\}.$$

It will be shown by using the theory in Chapter 8 that the equation (1.2) and then the equation (1.1), for $u_0 \in D(H(0))$, have a strong solution given by

$$u(t) = \lim_{n \to \infty} \prod_{i=1}^{n} I - \frac{t}{n}H(i\frac{t}{n})]^{-1}u_0$$

$$= \lim_{\nu \to 0} \prod_{i=1}^{[\frac{t}{\nu}]} I - \nu H(i\nu)]^{-1}u_0,$$

provided that certain condition of smallness is fulfilled.

The higher space dimensional analogue of (1.1) will be of the form

$$u_t(x,t) = \alpha_0(x,u,t,Du)\triangle u(x,t) + \tilde{g}_0(x,t,u,Du),$$
$$(x,t) \in \Omega \times (0,T);$$

$$\frac{\partial}{\partial \hat{n}}u(x,t) + \beta(x,t,u) = 0, \quad x \in \partial\Omega;$$
$$u(x,0) = u_0(x);$$

(1.3)

in which eight assumptions are made.

- Ω is a bounded smooth domain in $\mathbb{R}^N, N \geq 2$, and $\partial\Omega$ is the boundary of Ω.
- $\hat{n}(x)$ is the unit outer normal to $x \in \partial\Omega$, and μ is a real number such that $0 < \mu < 1$.
- $\alpha_0(x,t,z,p) \in C^{1+\mu}(\overline{\Omega} \times \mathbb{R} \times \mathbb{R}^N)$ is true for each $t \in [0,T]$ where $T > 0$, and is continuous in all its arguments. Furthermore, $\alpha_0(x,t,z,p) \geq \delta_1 > 0$ is true for all x,z,p, and all $t \in [0,T]$, and for some constant $\delta_1 > 0$.

- $\tilde{g}_0(x,t,z,p) \in C^{1+\mu}(\overline{\Omega} \times \mathbb{R} \times \mathbb{R}^N)$ is true for each $t \in [0,T]$, is continuous in all its arguments, and satisfies

$$z\tilde{g}_0(x,t,z,p) \le \delta_{11}|z|^2 + |z|b$$

for some real number δ_{11} and some positive number $b > 0$.

- $\tilde{g}_0(x,t,z,p)$ is of at most linear growth in p, that is,

$$|\tilde{g}_0(x,t,z,p)| \le M_{10}(x,t,z)(1+|p|)$$

for some nonnegative, continuous function M_{10} and for all $t \in [0,T]$.

- $\beta(x,t,z) \in C^{2+\mu}(\overline{\Omega} \times \mathbb{R})$ is true for each $t \in [0,T]$, is continuous in all its arguments, and is strictly monotone increasing in z so that $\beta_z \ge \delta_{11} > 0$ for the constant $\delta_{11} > 0$.

- The following are true for some continuous, nonnegative functions N_{10}, N_{20}, N_{30} and for some continuous function ζ of bounded variation:

$$|\alpha_0(x,t,z,p) - \alpha_0(x,\tau,z,p)|/\alpha_0(x,\tau,z,p)$$
$$\le |\zeta(t) - \zeta(\tau)|N_{10}(x,t,\tau,z);$$
$$|\tilde{g}_0(x,t,z,p) - \tilde{g}_0(x,\tau,z,p)| \le |\zeta(t) - \zeta(\tau)|N_{20}(x,t,\tau,z)(1+|p|);$$
$$|\beta(x,t,z) - \beta(x,\tau,z)| \le |t - \tau|N_{30}(x,t,\tau,z).$$

- The following are true for continuous, nonnegative functions N_{11} and N_{21}:

$$|\alpha_0(x,t,z_1,p) - \alpha_0(x,t,z_2,p)|/\alpha_0(x,\tau,z_2,p)$$
$$\le N_{11}(x,t,\tau,z_1,z_2)|z_1 - z_2|;$$
$$|\tilde{g}_0(x,t,z_1,p) - \tilde{g}_0(x,t,z_2,p)| \le N_{21}(x,t,z_1,z_2)(1+|p|)|z_1 - z_2|.$$

Again, it is to be noted that, relative to those in (1.3) of Chapter 6, here $\alpha_0(x,t,z,p)$ has the z dependence, and $\tilde{g}_0(x,t,z,p)$ need not be monotone non-increasing in z.

The corresponding nonlinear evolution equation will be

$$\frac{d}{dt}u(t) = J(t)u(t), \quad t \in (0,T);$$
$$u(0) = u_0, \tag{1.4}$$

in which the time-dependent, nonlinear operator

$$J(t) : D(J(t)) \subset C(\overline{\Omega}) \longrightarrow C(\overline{\Omega})$$

is defined by

$$J(t)v = \alpha_0(x,t,v,Dv)\triangle v + \tilde{g}_0(x,t,v,Dv);$$

$$v \in D(J(t)) \equiv \{w \in C^{2+\mu}(\overline{\Omega}) : \frac{\partial}{\partial \hat{n}}w + \beta(x,t,w) = 0 \quad \text{on } \partial\Omega\}.$$

The theory in Chapter 8 will be employed again to prove that, for $u_0 \in D(J(0))$, the quantity

$$u(t) = \lim_{n\to\infty} \prod_{i=1}^{n}[I - \frac{t}{n}J(i\frac{t}{n})]^{-1}u_0$$

$$= \lim_{\nu\to 0} \prod_{i=1}^{[\frac{t}{\nu}]}[I - \nu J(i\nu)]^{-1}u_0$$

is a strong solution for equation (1.4) or, equivalently, equation (1.3), as long as some condition of smallness is met.

The rest of this chapter is organized as follows. Section 2 states the main results, and Sections 3 and 4 prove the main results.

The material of this chapter is taken from our article [26].

2. Main Results

THEOREM 2.1. *The time-dependent operator $H(t)$ in (1.2) satisfies the four conditions, namely, the non-dissipativity condition (H3), the weaker range condition (H2)', the time-regulating condition (HC), and the embedding condition (HB) in Chapter 8. As a result, it follows from Theorems 2.1 and 2.3 of Chapter 8 that, for $u_0 \in D(H(0))$, the equation (1.2) and then the equation (1.1) have a strong solution given by the quantity*

$$u(t) = \lim_{n \to \infty} \prod_{i=1}^{n} [I - \frac{t}{n} H(i\frac{t}{n})]^{-1} u_0$$

$$= \lim_{\nu \to 0} \prod_{i=1}^{[\frac{t}{\nu}]} [I - \nu H(i\nu)]^{-1} u_0,$$

providing that the condition of smallness in Theorem 2.1 of Chapter 8 is satisfied, where involved are the Remark 2.2 in Chapter 8 and the (3.1) in Section 3. This condition of smallness includes the cases where sufficiently small is T or both $N_{03}(x,t,z_1,z_2)$ and $N_{02}(x,t,\tau,z_1,z_2)$ for finite $|z_1|$ and $|z_2|$, or both $H(0)u_0$ and the total variation of $\zeta(t)$ on $[0,T]$.

In that case, $u(t)$ also satisfies the middle equation in (1.1).

THEOREM 2.2. *The time-dependent operator $J(t)$ in (1.4) satisfies the four conditions, namely, the non-dissipativity condition (H3), the weaker range condition (H2)', the time-regulating condition (HC), and the embedding condition (HB) in Chapter 8. As a result, it follows from Theorems 2.1 and 2.3 of Chapter 8 that, for $u_0 \in D(J(0))$, the equation (1.4) and then the equation (1.3) have a strong solution given by the quantity*

$$u(t) = \lim_{n \to \infty} \prod_{i=1}^{n} [I - \frac{t}{n} J(i\frac{t}{n})]^{-1} u_0$$

$$= \lim_{\nu \to 0} \prod_{i=1}^{[\frac{t}{\nu}]} [I - \nu J(i\nu)]^{-1} u_0,$$

providing that the condition of smallness in Theorem 2.1 of Chapter 8 is satisfied, where involved are the Remark 2.2 in Chapter 8 and the (4.1) in Section 4. This condition of smallness includes the cases where small enough is T, or both $N_{21}(x,t,z_1,z_2)$ and $N_{11}(x,t,\tau,z_1,z_2)$ for finite $|z_1|$ and $|z_2|$, or, in addition to $N_{30}(x,t,\tau,z)$ for finite $|z|$, both $J(0)u_0$ and the total variation of $\zeta(t)$ on $[0,T]$.

In that case, $u(t)$ also satisfies the middle equation in (1.3).

3. Proof of One Space Dimensional Case

Proof of Theorem 2.1:

PROOF. We now begin the proof, which is composed of five steps.

Step 1. ($H(t)$ satisfies the non-dissipativity condition (H3).) The second condition in (H3) follows from Step 3 below, while the first condition in (H3) is done as in the proof of Theorem 2.1 of Chapter 4, where the maximum principle arguments were used.

Step 2. ($H(t)$ satisfies the weaker range condition $(H2)'$.) This is established by analogy with the proof of Theorem 2.1 in Chapter 4.

Step 3. ($H(t)$ satisfies the time-regulating condition (HC).) Let

$$g_i(x) \in C[0,1], \quad i = 1,2,$$

and let

$$v_1 = (I - \lambda H(t))^{-1} g_1;$$
$$v_2 = (I - \lambda H(\tau))^{-1} g_2;$$

where $\lambda > 0$ and $0 \le t, \tau \le T$. Then

$$(v_1 - v_2) - \lambda[\alpha(x,t,v_1,v_1')(v_1 - v_2)'' + \tilde{g}(x,t,v_1,v_1') - \tilde{g}(x,t,v_2,v_2')]$$
$$= \lambda\{\frac{[\alpha(x,t,v_1,v_1') - \alpha(x,t,v_2,v_1')] + [\alpha(x,t,v_2,v_1') - \alpha(x,\tau,v_2,v_2')]}{\alpha(x,\tau,v_2,v_2')}$$
$$[H(\tau)v_2 - \tilde{g}(x,\tau,v_2,v_2')] + [\tilde{g}(x,t,v_2,v_2') - \tilde{g}(x,\tau,v_2,v_2')]\}$$
$$+ (g_1 - g_2);$$
$$|\tilde{g}(x,t,v_1,v_2') - \tilde{g}(x,t,v_2,v_2')| \le N_{03}(x,t,v_1,v_2)(1 + |v_2'|)|v_1 - v_2|;$$
$$|\tilde{g}(x,t,v_2,v_2') - \tilde{g}(x,\tau,v_2,v_2')| \le |\zeta(t) - \zeta(\tau)|N_{01}(x,t,\tau,v_2)(1 + |v_2'|);$$
$$|\alpha(x,t,v_1,v_2') - \alpha(x,t,v_2,v_2')|/\alpha(x,\tau,v_2,v_2')$$
$$\le N_{02}(x,t,\tau,v_1,v_2)|v_1 - v_2|;$$
$$|\alpha(x,t,v_2,v_2') - \alpha(x,\tau,v_2,v_2')|/\alpha(x,\tau,v_2,v_2')$$
$$\le |\zeta(t) - \zeta(\tau)|N_{00}(x,t,\tau,v_2);$$
$$\|v_2''\|_\infty \le \delta_0^{-1}[\|H(\tau)v_2\|_\infty + \|\tilde{g}(x,\tau,v_2,v_2')\|_\infty];$$
$$H(\tau)v_2 = (v_2 - g_2)/\lambda = [(I - \lambda H(\tau))^{-1}g_2 - g_2]/\lambda;$$
$$|\tilde{g}(x,\tau,v_2,v_2')| \le M_{00}(x,\tau,v_2)(1 + |v_2'|);$$
$$(v_1 - v_2)'(0) \in [\beta_0(v_1(0)) - \beta_0(v_2(0))];$$
$$(v_1 - v_2)'(1) \in -[\beta_1(v_1(1)) - \beta_1(v_2(1))];$$

so there holds, for some functions $M_i, i = 1,2,3,4$, in the condition (HC),

$$\|v_1 - v_2\|_\infty \le [\|g_1 - g_2\|_\infty + \lambda M_1(\|v_1\|_\infty, \|v_2\|_\infty)]\|v_1 - v_2\|_\infty$$
$$+ \lambda M_2(\|v_1\|_\infty, \|v_2\|_\infty)(\|v_2 - g_2\|/\lambda)\|v_1 - v_2\|_\infty$$
$$+ \lambda|\zeta(t) - \zeta(\tau)|M_3(\|v_1\|_\infty, \|v_2\|_\infty)(\|v_2 - g_2\|/\lambda)$$
$$+ \lambda|\zeta(t) - \zeta(\tau)|M_4(\|v_1\|_\infty, \|v_2\|_\infty)]. \tag{3.1}$$

This is the condition (HC). Here used was the interpolation inequality [1], [13, page 135]

$$\|v_2'\|_\infty \le \epsilon \|v_2''\|_\infty + C(\epsilon)\|v_2\|_\infty$$

for each $\epsilon > 0$ and for some constant $C(\epsilon)$, where

$$\delta_0^{-1} \max_{x \in [0,1]; t \in [0,T]} |M_{00}(x,t,\|v_2\|_\infty)|\epsilon < 1$$

$$\text{if } \epsilon = 1/[\delta_0^{-1} \max_{x \in [0,1]; t \in [0,T]} |M_{00}(x,t,\|v_2\|_\infty)| + 2], \text{ for example.}$$

This is because, as in the proof of Theorem 2.1 of Chapter 4, the maximum principle applies; that is, there is an $x_0 \in (0,1)$ such that

$$\|v_1 - v_2\|_\infty = |(v_1 - v_2)(x_0)|;$$
$$(v_1 - v_2)'(x_0) = 0;$$
$$(v_1 - v_2)(x_0)(v_1 - v_2)''(x_0) \le 0.$$

Here $x_0 \in \{0,1\}$ is impossible, due to the boundary conditions.

Step 4. ($H(t)$ satisfies the embedding condition (HB) of embeddedly quasi-demi-closedness.) This follows as in the proof of Theorem 2.1 of Chapter 6.

Step 5. ($u(t)$ for $u_0 \in D(H(0))$ satisfies the middle equation in (1.1).) Consider the discretized equation

$$\begin{aligned} u_i - \nu H(t_i)u_i &= u_{i-1}, \\ u_i &\in D(H(t_i)), \end{aligned} \tag{3.2}$$

where $u_0 \in D(H(0)), i = 1, 2, \ldots, n, n \in \mathbb{N}$ is large, and $\nu > 0$ is such that

$$\nu\omega < 1 \text{ and } 0 \le t_i = i\nu \le T.$$

Here, for small enough ν,

$$u_i = \prod_{k=1}^{i} [I - \nu H(t_k)]^{-1} u_0$$

exists uniquely by Proposition 3.6 of Chapter 8. For convenience, we also define

$$u_{-1} = u_0 - \nu H(0)u_0.$$

Now, for each $t \in [0,T)$, we have $t \in [t_i, t_{i+1})$ for some i, so $i = [\frac{t}{\nu}]$. It follows from Theorem 2.1 of Chapter 8 that, for each above t with the corresponding i,

$$\lim_{\nu \to 0} u_i = \lim_{\nu \to 0} \prod_{k=1}^{[\frac{t}{\nu}]} [I - \nu H(t_k)]^{-1} u_0$$

$$= \lim_{n \to \infty} \prod_{k=1}^{n} [I - \frac{t}{n}H(k\frac{t}{n})]^{-1} u_0$$

$$\equiv u(t)$$

exists.

On the other hand, by utilizing Lemma 3.4 and Proposition 3.5 in Chapter 8, we have that $\|u_i\|_\infty$ and $\|H(t_i)u_i\|_\infty = \|(u_i - u_{i-1})/\nu\|_\infty$ are uniformly bounded. Those, in turn, result in a bound for $\|u_i\|_{C^2[0,1]}$ by the interpolation inequality [1], [13, page 135]. Therefore it follows from Ascoli-Arzela theorem [33] that a subsequence of u_i and then itself converge in $C^1[0,1]$ to a limit, as $\nu \longrightarrow 0$. This limit equals $u(t)$ as shown above. Consequently, $u(t)$ satisfies the middle equation in (1.1), as u_i does so.

The proof is complete. □

4. Proof of Higher Space Dimensional Case

Proof of Theorem 2.2:

PROOF. We now begin the proof, which consists of five steps.

Step 1. As in the proof of Theorem 2.2 of Chapter 4, the maximum principle arguments prove the first condition in (H3), whereas the second condition in (H3) is a result of Step 3 below. Here notice

$$\beta(x,t,z) = [\beta(x,t,z) - \beta(x,t,0)] + \beta(x,t,0)$$
$$= \beta_z(x,t,\theta z)(z-0) + \beta(x,t,0), \quad 0 \le \theta \le 1$$

by the mean value theorem, where $\beta_z \ge \delta_{11} > 0$.

Step 2. The proof of Theorem 2.2 in Chapter 4 also shows that $J(t)$ satisfies the weaker range condition $(H2)'$.

Step 3. ($J(t)$ satisfies the time-regulating condition (HC).) Let

$$g_i(x) \in C^\mu(\overline{\Omega}), \quad i = 1, 2,$$

and let

$$v_1 = (I - \lambda J(t))^{-1} g_1;$$
$$v_2 = (I - \lambda J(\tau))^{-1} g_2;$$

where $\lambda > 0$ and $0 \le t, \tau \le T$. Then

$$(v_1 - v_2) - \lambda[\alpha_0(x,t,v_1,Dv_1)\triangle(v_1 - v_2) + \tilde{g}_0(x,t,v_1,Dv_1) - \tilde{g}_0(x,t,v_2,Dv_2)]$$
$$= \lambda\{\frac{[\alpha_0(x,t,v_1,Dv_1) - \alpha_0(x,t,v_2,Dv_1)] + [\alpha_0(x,t,v_2,Dv_1) - \alpha_0(x,\tau,v_2,Dv_2)]}{\alpha_0(x,\tau,v_2,Dv_2)}$$
$$[J(\tau)v_2 - \tilde{g}_0(x,\tau,v_2,Dv_2)] + [\tilde{g}_0(x,t,v_2,Dv_2) - \tilde{g}_0(x,\tau,v_2,Dv_2)]\}$$
$$+ (g_1 - g_2), \quad x \in \Omega;$$
$$|\tilde{g}_0(x,t,v_1,Dv_2) - \tilde{g}_0(x,t,v_2,Dv_2)| \le N_{21}(x,t,v_1,v_2)(1 + |Dv_2|)|v_1 - v_2|;$$
$$|\tilde{g}_0(x,t,v_2,Dv_2) - \tilde{g}_0(x,\tau,v_2,Dv_2)| \le |\zeta(t) - \zeta(\tau)|N_{20}(x,t,\tau,v_2)(1 + |Dv_2|);$$
$$|\alpha_0(x,t,v_1,Dv_2) - \alpha_0(x,t,v_2,Dv_2)|/\alpha_0(x,\tau,v_2,Dv_2)$$
$$\le N_{11}(x,t,\tau,v_1,v_2)|v_1 - v_2|;$$
$$|\alpha_0(x,t,v_2,Dv_2) - \alpha_0(x,\tau,v_2,Dv_2)|/\alpha_0(x,\tau,v_2,Dv_2)$$
$$\le |\zeta(t) - \zeta(\tau)|N_{10}(x,t,\tau,v_2);$$
$$J(\tau)v_2 = (v_2 - g_2)/\lambda = [(I - \lambda J(\tau))^{-1}g_2 - g_2]/\lambda;$$
$$|\tilde{g}_0(x,\tau,v_2,Dv_2)| \le M_{10}(x,\tau,v_2)(1 + |Dv_2|);$$
$$\frac{\partial(v_1 - v_2)}{\partial\hat{n}} + [\beta(x,t,v_1) - \beta(x,t,v_2)]$$
$$= -[\beta(x,t,v_2) - \beta(x,\tau,v_2)], \quad x \in \partial\Omega;$$
$$|\beta(x,t,v_2) - \beta(x,\tau,v_2)| \le |t - \tau|N_{30}(x,t,\tau,v_2);$$

so there holds, for some functions $M_i, i = 0, 1, 2, 3, 4$, in the condition (HC),

$$
\begin{aligned}
\|v_1 - v_2\|_\infty \leq [\|g_1 - g_2\|_\infty &+ \lambda M_1(\|v_1\|_\infty, \|v_2\|_\infty)\|v_1 - v_2\|_\infty \\
&+ \lambda M_2(\|v_1\|_\infty, \|v_2\|_\infty)(\|v_2 - g_2\|/\lambda)\|v_1 - v_2\|_\infty \\
&+ \lambda|\zeta(t) - \zeta(\tau)|M_3(\|v_1\|_\infty, \|v_2\|_\infty)(\|v_2 - g_2\|/\lambda) \\
&+ \lambda|\zeta(t) - \zeta(\tau)|M_4(\|v_1\|_\infty, \|v_2\|_\infty)]; \quad \text{or}
\end{aligned}
\tag{4.1}
$$

$$\|v_1 - v_2\|_\infty \leq M_0(\|v_1\|_\infty, \|v_2\|_\infty)|t - \tau|;$$

proving the condition (HC). This is because, as in proving the non-dissipativity condition (H3) in Step 1, the maximum principle argument applies, that is, there is an $x_0 \in \overline{\Omega}$ such that

$$\|v_1 - v_2\|_\infty = |(v_1 - v_2)(x_0)|,$$

that, for $x_0 \in \Omega$,

$$D(v_1 - v_2)(x_0) = 0;$$
$$(v_1 - v_2)(x_0)\Delta(v_1 - v_2)(x_0) \leq 0,$$

and that, for $x_0 \in \partial\Omega$,

$$\frac{\partial(v_1 - v_2)}{\partial\hat{n}}(x_0) \geq 0 \quad \text{or} \leq 0 \text{ according as } (v_1 - v_2)(x_0) > 0 \text{ or } < 0.$$

Here, to derive, for some numbers $C_1(\|v_2\|_\infty)$ and $C_2(\|v_2\|_\infty)$, depending on $\|v_2\|_\infty$,

$$
\begin{aligned}
\|Dv_2\|_\infty &\leq \|v_2\|_{C^{1+\mu}(\overline{\Omega})} \\
&\leq C_1(\|v_2\|_\infty)\|\frac{v_2 - g_2}{\lambda}\|_\infty + C_2(\|v_2\|_\infty),
\end{aligned}
\tag{4.2}
$$

we used the integral representation of v_2, with the Green's function $Z(x, y)$ of the second kind [29],

$$
\begin{aligned}
v_2 = &- \int_{\partial\Omega} Z(x, y)[-\beta_2(y, \tau, v_2)] d\sigma_y \\
&+ \int_\Omega Z(x, y)\alpha_0(y, \tau, Dv_2)^{-1}[\frac{v_2 - g_2}{\lambda} - g_0(y, \tau, v_2, Dv_2)] dy.
\end{aligned}
$$

Indeed, the result follows from differentiating the above v_2 for $(1 + \mu)$ times, with respect to the variable x, to obtain, for some constants d_1 and d_2 and for some number $C_3(\|v_2\|_\infty)$ depending on $\|v_2\|_\infty$,

$$
\begin{aligned}
\|Dv_2\|_{C^\mu(\overline{\Omega})} \leq &d_1\|\frac{v_2 - g_2}{\lambda}\|_\infty \\
&+ d_2[\max_{x\in\overline{\Omega};t\in[0,T]} |M_{10}(x, t, \|v_2\|_\infty)|\|Dv_2\|_\infty + C_3(\|v_2\|_\infty)],
\end{aligned}
$$

and from combining the interpolation inequality [1], [13, page 135]

$$\|Dv_2\|_\infty \leq \epsilon\|Dv_2\|_{C^\mu(\overline{\Omega})} + C(\epsilon)\|v_2\|_\infty$$

for each $\epsilon > 0$ and for some constant $C(\epsilon)$. Here

$$\epsilon d_2 \max_{x\in\overline{\Omega};t\in[0,T]} |M_{10}(x, t, \|v_2\|_\infty)| < 1$$

$$\text{if } \epsilon = 1/[d_2 \max_{x\in\overline{\Omega};t\in[0,T]} |M_{10}(x, t, \|v_2\|_\infty)| + 2], \text{ for example,}$$

and, for some constants b_1 and b_2,

$$|D_i Z(x,y)| \leq b_1 |x-y|^{1-N};$$
$$|D_{ij} Z(x,y)| \leq b_2 |x-y|^{-N};$$
$$\int_\Omega |x-y|^{N(\mu-1)} \, dx \quad \text{is finite for } 0 < \mu < 1 \text{ [13, page 159]}.$$

Step 4. ($J(t)$ satisfies the embedding condition (HB) of embeddedly quasi-demi-closedness.) This follows as in the proof of Theorem 2.2 of Chapter 6.

Step 5. ($u(t)$ for $u_0 \in D(J(0))$ satisfies the middle equation in (1.3).) Consider the discretized equation

$$u_i - \nu J(t_i) u_i = u_{i-1},$$
$$u_i \in D(J(t_i)), \tag{4.3}$$

where $u_0 \in D(J(0)), i = 1, 2, \ldots, n$, $n \in \mathbb{N}$ is large, and $\nu > 0$ is such that

$$\nu\omega < 1 \text{ and } 0 \leq t_i = i\nu \leq T.$$

Here, for small enough ν,

$$u_i = \prod_{k=1}^{i} [I - \nu J(t_k)]^{-1} u_0$$

exists uniquely by Proposition 3.6 of Chapter 8. For convenience, we also define

$$u_{-1} = u_0 - \nu J(0) u_0.$$

Now, for each $t \in [0, T)$, we have $t \in [t_i, t_{i+1})$ for some i, so $i = [\frac{t}{\nu}]$. It follows from Theorem 2.1 of Chapter 8 that, for each above t with the corresponding i,

$$\lim_{\nu \to 0} u_i = \lim_{\nu \to 0} \prod_{k=1}^{[\frac{t}{\nu}]} [I - \nu J(t_k)]^{-1} u_0$$
$$= \lim_{n \to \infty} \prod_{k=1}^{n} [I - \frac{t}{n} J(k \frac{t}{n})]^{-1} u_0$$
$$\equiv u(t)$$

exists.

On the other hand, by utilizing Lemma 3.4 and Proposition 3.5 in Chapter 8, we have $\|u_i\|_\infty$ and $\|J(t_i) u_i\|_\infty = \|(u_i - u_{i-1})/\nu\|_\infty$ are uniformly bounded, whence so is $\|u_i\|_{C^{1+\lambda}(\overline{\Omega})}$ for any $0 < \lambda < 1$, using the proof of (4.2). (Alternatively, those, in turn, result in a bound for $\|u_i\|_{W^{2,p}(\Omega)}$ for any $p \geq 2$, by the L^p elliptic estimates [37]. Hence, a bound exists for $\|u_i\|_{C^{1+\eta}(\overline{\Omega})}$ for any $0 < \eta < 1$, as a result of the Sobolev embedding theorem [1, 13].) Therefore it follows from Ascoli-Arzela theorem [33] that a subsequence of u_i and then itself converge in $C^{1+\mu}(\overline{\Omega})$ to a limit, as $\nu \longrightarrow 0$. This limit equals $u(t)$ as shown above. Consequently, $u(t)$ satisfies the middle equation in (1.3), as u_i does so.

The proof is complete. $\qquad \square$

Appendix

In this Appendix, some essential background results from elliptic partial differential equations of second order will be collected for the convenience of the reader. There are five small sections in this Appendix.

1. Existence of a Solution

THEOREM 1.1 ([**5, 24**]). *The nonhomogeneous boundary value problem for the ordinary differential equations of second order*

$$p(x)y'' + q(x)y' + r(x)y = f(x), \quad a < x < b;$$

$$B\tilde{y} = \vec{\gamma} \in \mathbb{R}^2,$$

has a unique solution, if the homogeneous boundary value problem

$$p(x)y'' + q(x)y' + r(x)y = 0, \quad a < x < b,$$

$$B\tilde{y} = 0 \in \mathbb{R}^2,$$

only has the trivial zero solution.

Here

$$p(x), q(x), r(x), \quad \text{and } f(x) \text{ are continuous, real-valued}$$
$$\text{functions on } [a, b];$$

$$p(x) \geq \delta > 0 \quad \text{for some constant } \delta;$$

$$B \quad \text{is a real } 2 \times 4 \text{ matrix of rank 2};$$

$$\tilde{y} = \begin{pmatrix} y(a) \\ y'(a) \\ y(b) \\ y'(b) \end{pmatrix} \in \mathbb{R}^4 \quad \text{is a four dimensional, real vector;}$$

$$\vec{\gamma} \in \mathbb{R}^2 \quad \text{is a given two dimensional, real vector.}$$

Notice that the boundary conditions of the Robin, Neumann, Dirichlet or periodic type meet the requirements of Theorem 1.1.

THEOREM 1.2 ([**13**]). *The nonhomogeneous boundary value problem for the elliptic partial differential equation of second order with the Robin boundary condition*

$$Lu = \sum_{i,j=1}^{N} a_{ij}(x) D_{ij} u(x) + \sum_{i=1}^{N} b_i(x) D_i u(x)$$

$$+ c(x)u = f(x), \quad x \in \Omega \subset \mathbb{R}^N;$$

$$\gamma(x)u(x) + \beta(x) \frac{\partial u(x)}{\partial \nu} = \varphi(x), \quad x \in \partial\Omega,$$

has a unique solution in $C^{2+\alpha}(\overline{\Omega})$ *for all* $f \in C^{\alpha}(\overline{\Omega})$ *and* $\varphi \in C^{1+\alpha}(\overline{\Omega})$.

Here

$\qquad a_{ij}(x), b_i(x), c(x), \quad$ and $f(x)$ are in $C^{\alpha}(\overline{\Omega})$;

$\qquad \Omega \quad$ is a bounded smooth domain in $\mathbb{R}^N, N = 2, 3, \ldots$;

$\qquad \overline{\Omega} \quad$ is the closure of Ω;

$\qquad \partial\Omega \quad$ is the boundary of Ω;

$\qquad c(x) \leq 0 \quad$ for $x \in \overline{\Omega}$;

and

$\qquad L \quad$ is strictly elliptic, that is, the matrix function

$\qquad\qquad (a_{ij}(x))_{N \times N} \quad$ is strictly positive definite:

$$\sum_{i,j=1}^{N} a_{ij}(x)\xi_i\xi_j \geq \lambda|\xi| \quad \text{for some } \lambda > 0,$$

$$\text{for all } x \in \Omega \text{ and for all } \xi = \begin{pmatrix} \xi_1 \\ \vdots \\ \xi_N \end{pmatrix} \in \mathbb{R}^N;$$

$\qquad \gamma(x), \beta(x) \quad$ are in $C^{1+\alpha}(\overline{\Omega})$ with $\gamma(x)$ and $\beta(x)$

$\qquad\qquad$ positive on $\partial\Omega$ and $\beta(x) \geq \kappa > 0$ on $\partial\Omega$;

and

$\qquad \dfrac{\partial u(x)}{\partial \nu} \quad$ is the outer normal derivative of $u(x)$ on $\partial\Omega$;

$\qquad D_{ij}u(x) \quad$ are the second partial derivatives of $u(x)$;

$\qquad D_i u(x) \quad$ are the first partial derivatives of $u(x)$;

and

$\qquad C^k(\overline{\Omega}), k = 0, 1, \ldots, \quad$ is the real vector space of,

$\qquad\qquad k$ times, continuously differentiable, real-valued

$\qquad\qquad$ function on $\overline{\Omega}$;

$$\|u\|_{C^k(\overline{\Omega})} = \sum_{j=1}^{k} \sum_{|\beta|=k} \sup_{\overline{\Omega}} |D^{\eta}u(x)| \quad \text{for } u \in C^k(\overline{\Omega});$$

$$D^{\eta}u(x) = \frac{\partial^{|\eta|}u(x)}{\partial x_1^{\eta_1} \cdots \partial x_N^{\eta_N}}, x = (x_1, \ldots, x_N) \in \mathbb{R}^N;$$

$\qquad \eta = (\eta_1, \ldots, \eta_N), \eta_i \in \{0\} \cup \mathbb{N}, \quad$ is a multi-index

$$\text{with } |\eta| = \sum_{i=1}^{N} \eta_i;$$

and

$\qquad C^{\alpha}(\overline{\Omega}), 0 < \alpha < 1, \quad$ is the real vector space of all

$\qquad\qquad \alpha$-Holder continuous, real-valued functions on $\overline{\Omega}$;

$$\|u\|_{C^\alpha(\overline{\Omega})} = \sup_{x \neq y; x,y \in \overline{\Omega}} \frac{|u(x) - u(y)|}{|x - y|^\alpha} \quad \text{for } u \in C^\alpha(\overline{\Omega});$$

$$|x - y| = [\sum_{i=1}^{N} |x_i - y_i|^2]^{1/2} \quad \text{for } x = \begin{pmatrix} x_1 \\ \vdots \\ x_N \end{pmatrix},$$

$$y = \begin{pmatrix} y_1 \\ \vdots \\ y_N \end{pmatrix} \in \mathbb{R}^N;$$

and

$C^{k+\alpha}(\overline{\Omega})$ is the subspace of $C^k(\overline{\Omega})$ with functions

whose k-th order, partial derivatives are

are α-Holder continuous on $\overline{\Omega}$;

$$\|u\|_{C^{k+\alpha}(\overline{\Omega})} = \|u\|_{C^k(\overline{\Omega})} + \sup_{|\eta|=k} \|D^\eta u\|_{C^\alpha(\overline{\Omega})}.$$

THEOREM 1.3 ([**13**]). *Following Theorem 1.2, the nonhomogeneous boundary value problem for the elliptic partial differential equation of second order with the Dirichlet boundary condition*

$$Lu = \sum_{i,j=1}^{N} a_{ij}(x) D_{ij} u(x) + \sum_{i=1}^{N} b_i(x) D_i u(x)$$

$$+ c(x)u = f(x), \quad x \in \Omega \subset \mathbb{R}^N;$$

$$u(x) = \varphi(x), \quad x \in \partial\Omega,$$

has a unique solution in $C^{2+\alpha}(\overline{\Omega})$ for all $f \in C^\alpha(\overline{\Omega})$ and $\varphi \in C^{2+\alpha}(\overline{\Omega})$.

THEOREM 1.4 ([**13**], Contraction Mapping Principle). *If*

$$T : X \longrightarrow X$$

is a strict contraction, then T has a unique fixed point, that is,

$$Tx = x$$

for some unique $x \in X$.

Here

X is a real Banach space with the norm $\|\cdot\|$;

T is a strict contraction, if there is a number $0 < \theta < 1$, such that

$$\|Tx_1 - Tx_2\| \leq \theta \|x_1 - x_2\|$$

holds for $x_1, x_2 \in X$;

THEOREM 1.5 ([**13**], The Method of Continuity). *Let X and Y be two real Banach spaces with the norms $\|\cdot\|_X$ and $\|\cdot\|_Y$, respectively. Let*

$$L_0, L_1 : X \longrightarrow Y$$

be two linear bounded operators. Let, for each $t \in [0, 1]$,

$$L_t = (1 - t)L_0 + tL_1.$$

Supppose that there is a constant C, such that, for $x \in X$,

$$\|x\|_X \leq C\|L_t x\|_Y$$

holds for all $t \in [0, 1]$. Then L_1 is onto if and only if so is L_0.

PROOF. Assume for a moment that L_s is onto for some $s \in [0, 1]$. It will be shown that L_s is also one to one, from which L_s^{-1} exists. But this follows from the assumption that there is a constant C, such that, for $x \in X$,

$$\|x\|_X \leq C\|L_t x\|_Y$$

holds for all $t \in [0, 1]$.

By making use of $\mathcal{L}_s^{-1}$, the equation, for $y \in Y$ given,

$$L_\tau x = y$$

is equivalent to the equation

$$x = L_s^{-1}y + (\tau - s)L_s^{-1}(L_0 - L_1)x$$

from which a linear map

$$S : X \longrightarrow X,$$

$$Sx = S_s x \equiv L_s^{-1}y + (\tau - s)L_s^{-1}(L_0 - L_1)u$$

is defined. A fixed point x of $S = S_s$ will be a solution of the equation $L_\tau x = y$.

By choosing $\tau \in [0, 1]$ such that

$$|s - \tau| < \delta \equiv [C(\|L_0\|_{X \to Y} + \|L_1\|_{X \to Y})]^{-1},$$

it follows that $S = S_s$ is a strict contraction map. Therefore S has a unique fixed point by the Contraction Mapping Principle, Theorem 1.4. Hence L_τ is onto for τ satisfying $|\tau - s| < \delta$.

It follows that, by dividing $[0, 1]$ into subintervals of length less than δ and repeating the above arguments in a finite number of times, L_τ becomes onto for all $\tau \in [0, 1]$, provided that it is onto for some $\tau \in [0, 1]$. In particular, L_1 is onto if and only if so is L_0. □

2. Apriori Estimates

THEOREM 2.1 ([13]). *Following Theorem 1.2, let $u \in C^{2+\alpha}(\overline{\Omega})$ be a solution to the nonhomogeneous boundary value problem for the elliptic partial differential equation of second order with the Robin boundary condition*

$$Lu = \sum_{i,j=1}^{N} a_{ij}(x)D_{ij}u(x) + \sum_{i=1}^{N} b_i(x)D_i u(x)$$

$$+ c(x)u = f(x), \quad x \in \Omega \subset \mathbb{R}^N;$$

$$\gamma(x)u(x) + \beta(x)\frac{\partial u(x)}{\partial \nu} = \varphi(x), \quad x \in \partial\Omega.$$

Let

$$\|a_{ij}, b_i, c\|_{C^\alpha(\overline{\Omega})}, \quad \|\gamma, \beta\|_{C^{1+\alpha}(\overline{\Omega})} \leq \Lambda$$

holds for some constant Λ, where $i, j = 1, \ldots, N$.

Then the apriori estimate

$$\|u\|_{C^{2+\alpha}(\overline{\Omega})} \le C(\|u\|_{C^0(\overline{\Omega})} + \|f\|_{C^\alpha(\overline{\Omega})} + \|\varphi\|_{C^{1+\alpha}(\overline{\Omega})})$$

holds for some constant C, where

$$C = C(N, \alpha, \lambda, \Lambda, \kappa, \overline{\Omega})$$

depends on $N, \alpha, \ldots, \overline{\Omega}$.

THEOREM 2.2 ([**13**]). *Following Theorem 1.2, let $u \in C^{2+\alpha}(\overline{\Omega})$ be a solution to the nonhomogeneous boundary value problem for the elliptic partial differential equation of second order with the Dirichlet boundary condition*

$$Lu = \sum_{i,j=1}^{N} a_{ij}(x)D_{ij}u(x) + \sum_{i=1}^{N} b_i(x)D_i u(x)$$

$$+ c(x)u = f(x), \quad x \in \Omega \subset \mathbb{R}^N;$$

$$u(x) = \varphi(x), \quad x \in \partial\Omega.$$

Let

$$\|a_{ij}, b_i, c\|_{C^\alpha(\overline{\Omega})} \le \Lambda$$

holds for some constant Λ, where $i, j = 1, \ldots, N$.

Then the apriori estimate

$$\|u\|_{C^{2+\alpha}(\overline{\Omega})} \le C(\|u\|_{C^0(\overline{\Omega})} + \|f\|_{C^\alpha(\overline{\Omega})} + \|\varphi\|_{C^{2+\alpha}(\overline{\Omega})})$$

holds for some constant C, where

$$C = C(N, \alpha, \lambda, \Lambda, \overline{\Omega})$$

depends on $N, \alpha, \ldots, \overline{\Omega}$.

3. Hopf Boundary Point Lemma

LEMMA 3.1 ([**13**]). *Let u be a function defined on $\overline{\Omega}$, such that*

$$Lu = \sum_{i,j=1}^{N} a_{ij}(x)D_{ij}u(x) + \sum_{i=1}^{N} b_i(x)u(x)$$

$$+ c(x)u(x) \ge 0 \quad \text{for } x \in \Omega.$$

Suppose that, at a point $x_0 \in \partial\Omega$, $u(x_0)$ is a non-negative maximum. Then the outer, normal derivative $\frac{\partial u(x_0)}{\partial \nu}$ of u at x_0, if it exists, satisfies

$$\frac{\partial u(x_0)}{\partial \nu} > 0.$$

Here

Ω is a bounded smooth domain in $\mathbb{R}^N, N = 2, 3, \ldots$;

u is continuous at x_0;

L is uniformly elliptic, that is, the matrix function

$(a_{ij}(x))_{N \times N}$ satisfies the estimate:

$$0 < \lambda(x)|\xi|^2 \le \sum_{i,j=1}^{N} a_{ij}(x)\xi_i\xi_j \le \Lambda(x)|\xi|^2$$

for some functions $\lambda(x)$ and $\Lambda(x)$ with $\dfrac{\lambda(x)}{\Lambda(x)}$

bounded in Ω;

$c(x) \leq 0 \quad$ for $x \in \Omega$;

$\dfrac{b_i(x)}{\lambda(x)}, \dfrac{c(x)}{\lambda(x)} \quad$ are bounded in Ω.

4. Interpolation Inequality

LEMMA 4.1 ([13]). *Let Ω be a bounded smooth domain in $\mathbb{R}^N, N \geq 2$. Let*

$$j + \beta < k + \alpha,$$

where

$$0 < \alpha, \beta < 1; \quad j = 0, 1, \ldots; \quad k = 1, 2, \ldots.$$

Let $u \in C^{k+\alpha}(\overline{\Omega})$.

Then for any $\epsilon > 0$, there is a constant $C = C(\epsilon, j, k, \overline{\Omega})$, such that

$$\|u\|_{C^{j+\beta}(\overline{\Omega})} \leq C\|u\|_{C^0(\overline{\Omega})} + \epsilon\|u\|_{C^{k+\alpha}(\overline{\Omega})}$$

holds.

5. Sobolev Embedding Theorem

THEOREM 5.1 ([1]). *Let Ω be a bounded smooth domain in $\mathbb{R}^N, N \geq 2$. Let j and m be non-negative integers, and let p satisfy $1 \leq p < \infty$. Suppose that*

$$mp > N > (m - p).$$

Then the real Sobolev space $W^{j+m,p}(\Omega)$ is imbedded into the real vector space $C^{j+\lambda}(\overline{\Omega})$ for $0 < \lambda < m - \frac{N}{p}$. That is, for $u \in W^{j+m,p}(\Omega)$, there is a constant K, such that

$$\|u\|_{C^{j+\lambda}(\overline{\Omega})} \leq K\|u\|_{W^{j+m,p}(\Omega)}$$

holds.

Here the real Sobolev space $W^{k,p}(\Omega)$ for $k = 0, 1, 2, \ldots$, is defined as follows.

$W^{k,p}(\Omega) = \{u \in W^k(\Omega) : \text{the weak derivatives}$

$D^\eta u \in L^p(\Omega)$ for all multi-indices η with $|\eta| \leq k\}$;

$W^k(\Omega) = \{u : u$ is a k times, weakly differentiable,

real-valued function on $\Omega\}$;

$L^p(\Omega) = \{u : u$ is a measurable, real-valued

function on Ω that is p-integrable$\}$.

Bibliography

[1] R. A. Adams, *Sobolev Spaces*, Academic Press, New York, 1975.

[2] T. M. Apostol, *Mathematical Analysis*, second edition, Addison-Wesley Publishing Company, Inc., 1974.

[3] V. Barbu, *Semigroups and Differential Equations in Banach Spaces*, Leyden: Noordhoff, 1976.

[4] C.-C. Chen and K.-M. Koh, *Principles and Techniques in Combinatorics*, World Scientific, Singapore, 1992.

[5] E. A. Coddington and N. Levinson, *Theory of Ordinary Differential Equations*, McGraw-Hill Book Company Inc., New York, 1955.

[6] M. G. Crandall and T. M. Liggett, *Generation of semigroups of nonlinear transformations on general Banach spaces*, Amer. J. Math., **93** (1971), 256-298.

[7] M. G. Crandall, *A generalized domain for semigroup Generators*, Proceedings of the AMS, **2**, (1973), 435-440.

[8] M. G. Crandall and A. Pazy, *Nonlinear evolution equations in Banach spaces*, Israel J. Math., **11** (1972), 57-94.

[9] K. Engel and R. Nagel, *One-Parameter Semigroups for Linear Evolution Equations*, Springer-Verlag, New York, 1999.

[10] K. Engel and R. Nagel, *A Short Course on Operator semigroups*, Springer-Verlag, New York, 2006.

[11] L. Gaul, M. Kogl, and M. Wagner, *Boundary Element Methods for Engineers And Scientists: An Introductory Course with Advanced Topics*, Springer-Verlag, New York, 2003.

[12] J. Kacur, *Method of Rothe in Evolution Equations*, Teubner Texte Zur Mathematik, Band bf80, BSB B. G. Teubner Verlagsgessellschaft, Leipzig, 1985.

[13] D. Gilbarg and N. S. Trudinger, *Elliptic Partial Differential Equations of Second Order*, Second Edition, Springer-Verlag, New York, 1983.

[14] J. A. Goldstein, *Semigroups of Linear Operators and Applications*, Oxford University Press, New York, 1985.

[15] J. A. Goldsetin and C.-Y. Lin, *Singular nonlinear parabolic boundary value problems in one space dimension*, J. Diff. Eqns., **68** (1987), 429-443.

[16] E. Hille and R. S. Phillips, *Functional Analysis and Semi-groups*, Amer. Math. Soc. Coll. Publ., Vol. 31, Providence, R. I., 1957.

[17] T. Kato, *Perturbation Theory for Linear Operators*, Springer, New York, 1966.

[18] C.-Y. Lin, *Degenerate nonlinear parabolic boundary value problems*, Nonlinear Analysis, T. M. A., **13** (1989), 1303-1315.

[19] C.-Y. Lin, *Quasilinear parabolic equations with nonlinear boundary conditions*, J. Computational and Applied Math., **126** (2000), 339-349.

[20] C.-Y. Lin, *Cauchy problems and applications*, Topological Methods in Nonlinear Analysis, **15** (2000), 359-368.

[21] C.-Y. Lin, *On generation of C_0 semigroups and nonlinear operator semigroups*, Semigroup Forum, **66** (2003), 110-120.

[22] C.-Y. Lin, *On generation of nonlinear operator semigroups and nonlinear evolution operators*, Semigroup Forum, **67** (2003), 226-246.

[23] C.-Y. Lin, *Nonlinear evolution equations*, Electronic Journal of Differential Equations, Vol. 2005 (2005), No. 42, pp. 1-42.

[24] C.-Y. Lin, *Theory and Examples of Ordinary Differential Equations*, World Scientific, Singapore, 2011.

[25] C.-Y. Lin, *Time-dependent domains for nonlinear evolution operators and partial differential equations*, Electronic Journal of Differential Equations, Vol. 2011 (2011), No. 92, pp. 1-30.

[26] C.-Y. Lin, *Some non-dissipativity condition for evolution equations*, Int. J. Math., Vol. 24, No. 2 (2013), 1350002 (28 pages).

[27] C. L. Liu, *Introduction to Combinatorial Mathematics*, McGraw-Hill, New York, 1968.

[28] R. E. Mickens, *Difference Equations, Theory and Applications*, Second Edition, Van Mostrand Reinhold, New York, (1990).

[29] C. Miranda, *Partial Differential Equations of Elliptic Type*, Springer-Verlag, New York, 1970.

[30] I. Miyadera, *Nonlinear Semigroups*, Translations of Mathematical Monographs, Vol. 109, American Mathematical Society, 1992.

[31] A. Pazy, *Semigroups of Linear Operators and Applications in Partial Differential Equations*, Springer-Verlag, New York, 1983.

[32] E. Rothe, *Zweidimensionale parabolische Randvertaufgaben als Grenfall eindimensionale Renvertaufgaben*, Math. Ann., **102** (1930), 650-670.

[33] H. L. Royden, *Real Analysis*, Macmillan Publishing Company, New York, 1989.

[34] A. Schatz, V. Thomee, and W. Wendland, *Mathematical Theory of Finite and Boundary Element Methods*, Birkhauser, Basel, Boston, 1990.

[35] H. Serizawa, *M-Browder-accretiveness of a quasi-linear differential operator*, Houston J. Math., **10** (1984), 147-152.

[36] A. E. Taylor and D. C. Lay, *Introduction to Functional Analysis*, Wiley, New York, (1980).

[37] G. M. Troianiello, *Elliptic Differential Equations and Obstacle Problems*, Plenum Press, New York, 1987.

[38] U. Westphal, *Sur la saturation pour des semi-groups ono lineaires*, C. R. Acad. Sc. Paris **274** (1972), 1351-1353.

[39] K. Yosida, *Functional Analysis*, Springer, New York, 1980.

Index

Printed in the United States
By Bookmasters